Game Theory for Data Science

Eliciting Truthful Information

Synthesis Lectures on Artificial Intelligence and Machine Learning

Editors
Ronald J. Brachman, ***Jacobs Technion-Cornell Institute at Cornell Tech***
Peter Stone, ***University of Texas at Austin***

Game Theory for Data Science: Eliciting Truthful Information
Boi Faltings and Goran Radanovic
2017

Multi-Objective Decision Making
Diederik M. Roijers and Shimon Whiteson
2017

Lifelong Machine Learning
Zhiyuan Chen and Bing Liu
2016

Statistical Relational Artificial Intelligence: Logic, Probability, and Computation
Luc De Raedt, Kristian Kersting, Sriraam Natarajan, and David Poole
2016

Representing and Reasoning with Qualitative Preferences: Tools and Applications
Ganesh Ram Santhanam, Samik Basu, and Vasant Honavar
2016

Metric Learning
Aurélien Bellet, Amaury Habrard, and Marc Sebban
2015

Graph-Based Semi-Supervised Learning
Amarnag Subramanya and Partha Pratim Talukdar
2014

Robot Learning from Human Teachers
Sonia Chernova and Andrea L. Thomaz
2014

General Game Playing
Michael Genesereth and Michael Thielscher
2014

Judgment Aggregation: A Primer
Davide Grossi and Gabriella Pigozzi
2014

An Introduction to Constraint-Based Temporal Reasoning
Roman Barták, Robert A. Morris, and K. Brent Venable
2014

Reasoning with Probabilistic and Deterministic Graphical Models: Exact Algorithms
Rina Dechter
2013

Introduction to Intelligent Systems in Traffic and Transportation
Ana L.C. Bazzan and Franziska Klügl
2013

A Concise Introduction to Models and Methods for Automated Planning
Hector Geffner and Blai Bonet
2013

Essential Principles for Autonomous Robotics
Henry Hexmoor
2013

Case-Based Reasoning: A Concise Introduction
Beatriz López
2013

Answer Set Solving in Practice
Martin Gebser, Roland Kaminski, Benjamin Kaufmann, and Torsten Schaub
2012

Planning with Markov Decision Processes: An AI Perspective
Mausam and Andrey Kolobov
2012

Active Learning
Burr Settles
2012

Computational Aspects of Cooperative Game Theory
Georgios Chalkiadakis, Edith Elkind, and Michael Wooldridge
2011

Representations and Techniques for 3D Object Recognition and Scene Interpretation
Derek Hoiem and Silvio Savarese
2011

A Short Introduction to Preferences: Between Artificial Intelligence and Social Choice
Francesca Rossi, Kristen Brent Venable, and Toby Walsh
2011

Human Computation
Edith Law and Luis von Ahn
2011

Trading Agents
Michael P. Wellman
2011

Visual Object Recognition
Kristen Grauman and Bastian Leibe
2011

Learning with Support Vector Machines
Colin Campbell and Yiming Ying
2011

Algorithms for Reinforcement Learning
Csaba Szepesvári
2010

Data Integration: The Relational Logic Approach
Michael Genesereth
2010

Markov Logic: An Interface Layer for Artificial Intelligence
Pedro Domingos and Daniel Lowd
2009

Introduction to Semi-Supervised Learning
XiaojinZhu and Andrew B.Goldberg
2009

Action Programming Languages
Michael Thielscher
2008

Representation Discovery using Harmonic Analysis
Sridhar Mahadevan
2008

Essentials of Game Theory: A Concise Multidisciplinary Introduction
Kevin Leyton-Brown and Yoav Shoham
2008

A Concise Introduction to Multiagent Systems and Distributed Artificial Intelligence
Nikos Vlassis
2007

Intelligent Autonomous Robotics: A Robot Soccer Case Study
Peter Stone
2007

Game Theory for Data Science: Eliciting Truthful Information

Boi Faltings and Goran Radanovic

www.morganclaypool.com

ISBN: 9781627057295 paperback
ISBN: 9781627056083 ebook

DOI 10.2200/S00788ED1V01Y201707AIM035

A Publication in the Morgan & Claypool Publishers series
SYNTHESIS LECTURES ON ARTIFICIAL INTELLIGENCE AND MACHINE LEARNING

Lecture #35
Series Editors: Ronald J. Brachman, *Jacobs Technion-Cornell Institute at Cornell Tech*
Peter Stone, *University of Texas at Austin*
Series ISSN
Print 1939-4608 Electronic 1939-4616

Game Theory for Data Science

Eliciting Truthful Information

Boi Faltings
École Polytechnique Fédérale de Lausanne (EPFL)

Goran Radanovic
Harvard University

SYNTHESIS LECTURES ON ARTIFICIAL INTELLIGENCE AND MACHINE LEARNING #35

MORGAN & CLAYPOOL PUBLISHERS

ABSTRACT

Intelligent systems often depend on data provided by information agents, for example, sensor data or crowdsourced human computation. Providing accurate and relevant data requires costly effort that agents may not always be willing to provide. Thus, it becomes important not only to verify the correctness of data, but also to provide incentives so that agents that provide high-quality data are rewarded while those that do not are discouraged by low rewards.

We cover different settings and the assumptions they admit, including sensing, human computation, peer grading, reviews, and predictions. We survey different incentive mechanisms, including proper scoring rules, prediction markets and peer prediction, Bayesian Truth Serum, Peer Truth Serum, Correlated Agreement, and the settings where each of them would be suitable. As an alternative, we also consider reputation mechanisms. We complement the game-theoretic analysis with practical examples of applications in prediction platforms, community sensing, and peer grading.

KEYWORDS

data science, information elicitation, multi-agent systems, computational game theory, machine learning

Contents

Preface

Data has very different characteristics from material objects: its value is crucially dependent on novelty and accuracy, which are determined only from the context where it is generated. On the other hand, it can be freely copied at no extra cost. Thus, it cannot be treated as a resource with an intrinsic value, as is the focus in most of game theory.

Instead, we believe that game theory for data has to focus on incentives for *generating* novel and accurate data, and we bring together a body of recent work that takes this perspective.

We describe a variety of mechanisms that can be used to provide such incentives. We start by showing incentive mechanisms for verifiable information, where a ground truth can be used as a basis for incentives. Most of this book is about the much harder problem of incentives for unverifiable information, where the ground truth is never known. It turns out that even in this case, game-theoretic schemes can provide incentives that make providing accurate and truthful information the best interest of contributors.

We also consider scenarios where agents are mainly interested in influencing the result of learning algorithms through the data they provide, including malicious agents that do not respond to monetary rewards. We show how the negative influence of any individual data provider on learning outcomes can be limited and thus how to thwart malicious reports.

While our main goal is to make the reader understand the principles for constructing incentive mechanisms, we finish by addressing several other aspects that have to be considered for their integration in a practical distributed machine learning system.

This book is a snapshot of the state of the art in this evolving field at the time of this writing. We hope that it will stimulate interest for further research, and make it itself obsolete soon!

Boi Faltings and Goran Radanovic
July 2017

Acknowledgments

Our interest in this topic goes back to 2003 and much of the early work was carried out in collaboration with Radu Jurca, who has developed several of the mechanisms described in this book and is responsible for many important insights. We also thank numerous researchers for discussions and comments over the years, in particular Yiling Chen, Vincent Conitzer, Chris Dellarocas, Arpita Ghosh, Kate Larson, David Parkes, David Pennock, Paul Resnick, Tuomas Sandholm, Mike Wellmann, and Jens Witkowski.

Boi Faltings and Goran Radanovic
July 2017

CHAPTER 1

Introduction

1.1 MOTIVATION

The use of *Big Data* for gaining insights or making optimal decisions has become a mantra of our time. There are domains where data has been used for a long time, such as financial markets or medicine, and where the improvements in current technology have enabled innovations such as automated trading and personalized health. The use of data has spread to other domains, such as choosing restaurants or hotels based on reviews by other customers and one's own past preferences. Automated recommendation systems have proven very useful in online dating services and thus influence the most important choice of our lives, that of our spouse. More controversial uses, such as to profile potential terrorists, profoundly influence our society already today.

Given that data is becoming so important—it has been called the "oil of the 21st century"—it should not be restricted to be used only by the entity that collected it, but become a commodity that can be traded and shared. Data science will become a much more powerful tool when organizations can gather and combine data gathered by others, and outsource data collection to those that can most easily observe it.

However, different from oil, it is very difficult to tell the *quality* of data. Just from a piece of data itself, it is impossible to tell a random number from an actual measurement. Furthermore, data may be redundant with other data that is already known. Clearly, the quality of data depends on its context, and paying for data according to its quality will require more complex schemes than for oil.

Another peculiarity of data is that it can be copied for free. A value is generated only when data is observed for the first time. Thus, it makes sense to focus on how to reward those that provide those initial observations in a way that not only compensates them for the effort, but motivates them to provide the best possible quality. This is the problem we address in this book.

To understand the quality issue, let us consider four examples where data is obtained from others: product reviews, opinion polls, crowdsensing, and crowdwork.

1.1.1 EXAMPLE: PRODUCT REVIEWS

Anyone who is buying a product, choosing a restaurant, or booking a hotel should take into account the experiences of other, like-minded customers with the different options. Thus, *reviews* have become one of the biggest successes of the internet, a case where users truly share information for each others' benefit. Reviews today are so essential for running a restaurant or a hotel that there is a lot of reason for manipulating them, and we have to wonder if we can trust

Figure 1.1: Customer reviews.

any of this information. While review sites are going to great lengths to eliminate actual fake reviews that have been written for money, there is still a self-selection bias because reviews are written voluntarily, as we shall see below.

Who would have any motivation to leave a review, given that writing it takes effort and there is absolutely no reward. While there are some truly altruistic reviewers, many of them fall into two categories: those that are extremely unhappy and want to "get even," and those that have been treated extremely well, often with the hint that a nice review would be welcome. This tendency is more than anecdotal: Hu, Pavlou, and Zhang [1] report a detailed analysis of reviews found on the `Amazon.com` website and show that most of them have a skewed distribution like the one shown in Figure 1.2. However, when they asked an unbiased population of 66 students to test the product whose review distribution is shown on the left, they obtained a distribution close to a Gaussian, as shown on the right in Figure 1.2. The same authors have further extended their study and shown on four different product categories that voluntary review distributions often contain strong biases, and furthermore argue that these biases cannot be easily corrected [2].

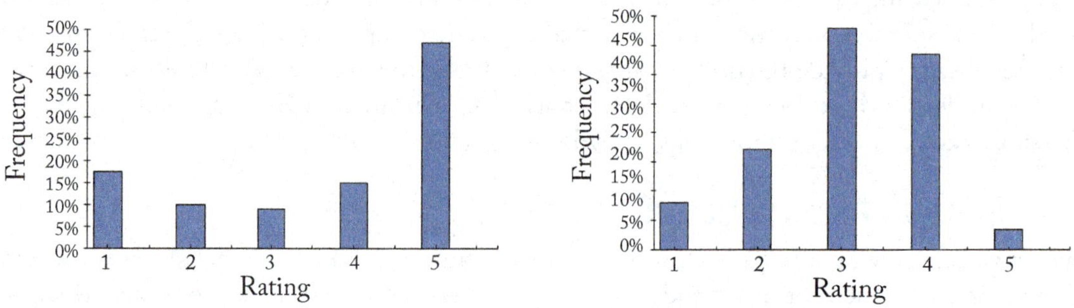

Figure 1.2: Distribution of review scores for a particular music CD on `Amazon.com` and among all students in a class (as reported in Hu, Pavlou, and Zhang [1]).

Given such a biased distribution, we have to ask ourselves if it makes sense to give so much weight to reviews when we make decisions. Clearly, there is a lot of room for improvement in the way that reviews are collected, and this will be one of the main topics of this book.

1.1.2 EXAMPLE: FORECASTING POLLS

Another area where data has a huge impact on our lives are opinion polls, such as forecasts of election outcomes, of public opinion about policies and issues, or of the potential success of new products. We have seen some spectacular failures of these polls, such as in predicting the outcome of the vote about Scottish independence in 2014 (see Figure 1.3), or the vote about Britain leaving the EU in 2016. Political campaigns and party platforms could more accurately reflect true voter preferences if they could rely on accurate polls. But political polls are just the tip of the iceberg: huge inefficiencies are caused by products and services that do not fit the needs of their customers because these have not been accurately determined by the polls that were used in planning.

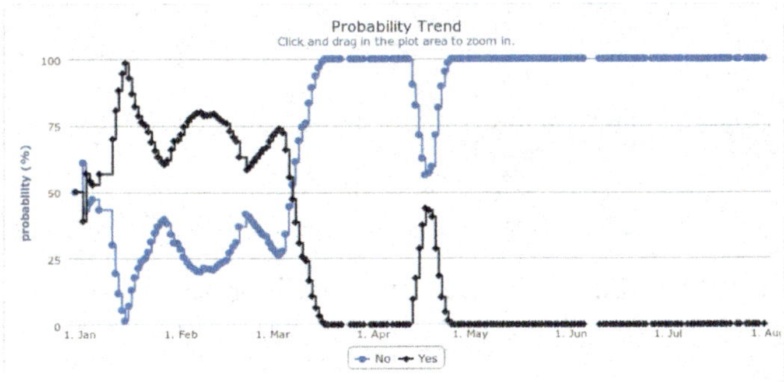

Figure 1.3: Evolution of the online opinion poll `Swissnoise` during the time leading up the referendum on Scottish independence on September 18, 2014.

The internet could provide a great tool for collecting this information, but we find similar issues of self-selection and motivation as we had with reviews. People who are fans of particular idols or who have particular agendas will go to great lengths to influence poll outcomes to serve their ulterior motives, while the average citizen has no interest in spending the effort necessary to answer polls. We need schemes that encourage unbiased, knowledgeable participants, and accurate estimates.

1.1.3 EXAMPLE: COMMUNITY SENSING

According to an estimate released by the World Health Organization in March 2014, air pollution kills 7 million people each year. This makes it a more serious issue than traffic accidents, which are held responsible for 1.25 million deaths by the same organization. While it was long thought that air pollution spreads quite uniformly across cities, recent research has shown it to be a very localized phenomenon with big variations of over 100% even in the space of 100 m. Figure 1.4 shows an example of fine particle distribution in the city of Beijing. Huge variations of up to a factor of 5 exist not only between places that are only a few kilometers apart, but also at one and the same place within just one hour. The influence on human health could be reduced if people could adjust their movements to minimize exposure to the most polluted zones.

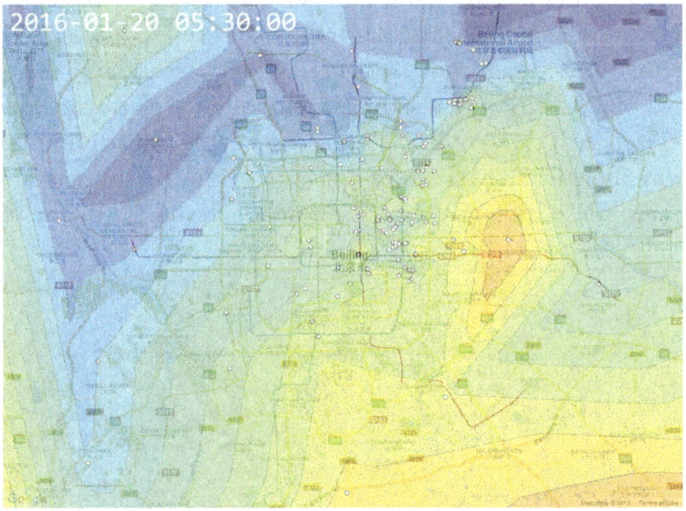

Figure 1.4: Fine particle distribution in the city of Beijing. Courtesy of Liam Bates, Origins Technology Limited.

While some pollution can be seen and smelled, truly harmful substances such as fine particles, CO and NO_2, cannot be detected by humans, so they must be alerted by sensors to avoid exposure. Newly developed low-cost sensors hold the promise to obtain real-time measurements at reasonably low cost. As a city does not have easy access to all locations, the best way to deploy such sensors is through community sensing, where sensors are owned and operated by individuals and they get paid by the city for feeding the measurements into a system that aggregates them into a pollution map that is made available to the public.

An early example of such a low-cost sensor is the Air Quality Egg, an open source design developed in a kickstarter project in 2013, and sold over 1,000 times at a price of $185 (2013 price). The measurements are uploaded to a center controlled by manufacturer, and thus provide a (albeit not dense enough) map of air pollution that everyone can inspect. While the accuracy of such first-generation sensors is insufficient for practical use, the strong public interest shows that such schemes are feasible.

Sensor quality is improving rapidly. At the time of this writing, the LaserEgg shown in Figure 1.5 offers highly accurate fine particle measurements for just $85 (2016 price), and more comprehensive devices are under development at similar low cost. Buyers of the LaserEgg can upload their data to the internet, and the map in Figure 1.5 shows the distribution of the thousands of devices that have been connected in the city of Beijing alone; together they produce a the real-time pollution map shown in Figure 1.4.

Figure 1.5: Example of a low-cost sensor: the Laser Egg. Courtesy of Liam Bates, Origins Technology Limited.

Thus, we can expect community sensing to become commonplace in the near future. However, an important issue is how to compensate sensor operators for their efforts in procuring and operating the sensors. This problem will be solved by the techniques we show in this book.

1.1.4 EXAMPLE: CROWDWORK

It has become common to outsource intellectual work, such as programming, writing, and other tasks, through the internet. An extreme form of such outsourcing is *crowdsourcing*, where small tasks are given to a large anonymous *crowd* of workers who get small payments for executing each task. Examples of tasks found on the Amazon Mechanical Turk platform shown in Figure 1.6 include verifying the consistency of information on websites, and labeling images or natural language texts.

Crowdwork is also used in massive open online courses (MOOC) where students grade each others' homework, a process known as *peer grading*.

Later in the book, we will show practical experiences with incentive schemes both in crowdwork and peer grading. The difficulty in crowdwork is not only to obtain the right selection of workers, but also to make them invest the *effort* needed to obtain results of good quality. Thus, it is important that the incentives are significant enough to cover the cost of this effort.

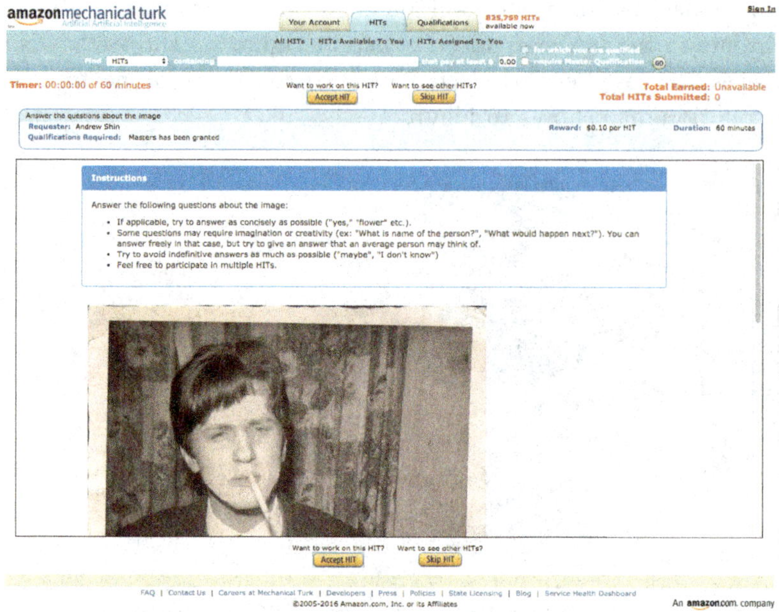

Figure 1.6: Example of a crowdworking platform: Amazon Mechanical Turk.

1.2 QUALITY CONTROL

There are three different ways to improve the quality of contributed data. They can all be used together, and in fact often they can mutually strengthen each other.

The first and easiest to implement is *filtering*: eliminating outliers and data that is otherwise statistically inconsistent. For example, in crowdwork it is common to give the same task to several workers, and apply a weighted majority vote to determine the answer. When noise and biases are known in more detail, statistical methods, such as those by Dawid and Skene [5], can help to the underlying truth. As these techniques are well developed, they are not our focus in this book, and should be used in parallel with the techniques we show.

The second way is to associate a quality score to the agents providing the data. For example, in a crowdworking setting it is common to include *gold* tasks that have known answers; workers are assigned a quality score based on the proportion of gold tasks they answer correctly. Alternatively, worker quality can be estimated as a latent variable [3] by fitting a model of both worker accuracy and true values to the reported data.

The third way is to provide *incentives* for agents to do their best to provide accurate information. For example, in a prediction platform, we could reward data as a function of how accurately it predicted the eventual outcome. However, in many cases, data is not that easy to verify, either because it may be subjective, as in customer experience, or the ground truth may never be known, as in most elicitation settings. Somewhat surprisingly, using game-theoretic mechanisms it is nevertheless possible to provide strong incentives for accurate data in most practical settings, and this is the focus of this book.

Among the three options, incentives is the only one that does not need to throw away data—in fact, they work to increase the amount and quality of available data. Incentives can be somewhat inaccurate, as long as agents *believe* that they are correct on average. Even a slightly inaccurate incentive can elicit correct data 100% of the time, while filtering and reputation can never completely eliminate inaccuracies.

However, an important condition for incentives to work is that the agents providing the information are *rational*: they act to optimize their expected reward. Agents who misunderstand, who don't care, or who have strong ulterior motives will not react. Thus, it is important for the schemes to be *simple* and easy to evaluate by agents providing information.

1.3 SETTING

In this book, we consider the multi-agent setting shown in Figure 1.7. We collect data about the state of a phenomenon, which can be, for example, the quality of a restaurant in the case of reviews, the true answer to a task in the case of crowdwork, or a physical phenomenon such as the pollution in a city.

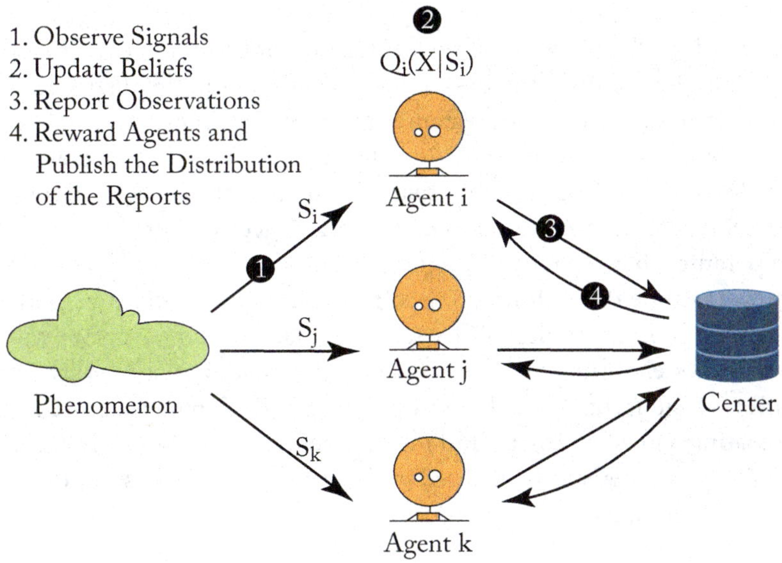

Figure 1.7: Multi-agent setting assumed in this book.

The state is characterized by a variable X that takes values in a discrete space $\{x_1, \ldots, x_n\}$. The phenomenon is observed by several *information agents* $A = \{a_1, \ldots, a_k\}$. Agent a_i observes a *signal* s_i taken from the same vocabulary as the states so that we assume that s_i takes values in $\{x_1, \ldots, x_n\}$ as well.

A *center* is interested in data about the phenomenon in order to learn a model or make decisions that depend on it. It asks the agents to report their information about the phenomenon. In return, it provides agents with a *reward* that is chosen to motivate the agents to provide the highest quality of information possible.

We distinguish between *objective* data, where every agent observes the *same* realization of the phenomenon, and *subjective* data, where each agent observes a possibly *different* realization of the same phenomenon. Measuring the temperature at a particular time and place is an example of objective data. Judging the quality of meals in a restaurant is subjective data since every agent eats a different meal but from the same kitchen. For objective data, the center is interested in obtaining the most accurate estimate of this value, whereas for subjective data, the goal will usually be to obtain the *distribution* of values observed by different agents.

For objective data, the state of the phenomenon has a ground truth that distinguishes accurate from inaccurate reports. We distinguish the case of *verifiable* information, where the center will eventually have access to the ground truth, and *unverifiable* information, where such ground truth will never be known or does not exist. Examples of verifiable information are weather forecasts and election polls. Subjective data such as a product review is always unverifiable; and in practice most objective data is also never verified since it would be too costly to do so.

For subjective data, the objective of the center cannot be to obtain a ground truth, as such a ground truth cannot be defined. The best it can do is to model the distribution of the signals observed by the information agents: for restaurant reviews, the center would like to predict how much another customer will like the meal, not the reasons why the restaurant obtains meals with this quality distribution. Even for objective information, the center will sometimes be more interested in modeling the observations rather than the objective truth: weather forecasts report a subjective temperature that accounts for wind and humidity.

Therefore, we assume throughout this book that the center's objective is to obtain accurate reports of the *signals* observed by the information agents. Consequently, we will consider a superficial but simple model of the phenomenon where the state is just the distribution of signals that the population of agents observes. For example, for pollution measurements this would be a distribution indicating a noisy estimate, and for product reviews it would reflect the distribution of ratings. This allows methods to be general without the need for assumptions about detailed models of the phenomenon itself.

Influencing an agent's choice of strategy The crucial elements of the scenario, as given above, are the agents that observe the phenomenon. Each agent is free to choose a *strategy* for reporting this observation to the center. We distinguish *heuristic* strategies, where the agent does not even make the effort to observe the phenomenon.

Definition 1.1 A reporting strategy is called *heuristic* when the reported value does not depend on an observation of the phenomenon.

And *cooperative* strategies.

Definition 1.2 A reporting strategy is called *cooperative* if the agent invests effort to observe the phenomenon and truthfully reports the observation.

Examples of heuristic strategies are to always report the same data, to report random data, or to report the most likely value according to the prior distribution. In cooperative strategies, we sometimes distinguish different levels of effort that result in different accuracy. Cooperative strategies are a subclass of *truthful* strategies, where agents report their belief about the phenomenon truthfully. We will see later that when we are able to strictly incentivize truthful strategies, we can also incentivize the effort required for cooperative strategies with the proper scaling.

Except in the chapter on limiting influence, we assume that the agents have no interest in influencing the data collected by the center toward a certain result. We assume furthermore that agents are *rational* and risk-neutral in that they always chose the strategy that maximizes their *expected utility*, and we assume that their utilities are quasi-linear so that they can be calculated as the difference between reward and cost of their effort.

These characteristics make it possible for the center to influence the agents' choice of strategy through the rewards they offer. In game theory, this is called a *mechanism*, and the design of such mechanisms is the main topic of this book. In general, we strive to achieve the following properties:

- truthfulness: it induces agents to choose a cooperative and truthful strategy;
- individually rational: agents can expect a positive utility from participating; and
- positive self-selection: only agents that are capable of providing useful data can expect a positive utility from participating.

Principle for mechanism design Agents incur a cost for obtaining and reporting data to the center. They will not do so without being at least compensated for this cost. Depending on the circumstances, they may be interested in *monetary* compensations for the cost they incur (where monetary could also mean recognition, badges, or other such rewards), or they may be interested in *influencing* the model the center learns from their data. In most of the book, we consider the case of monetary incentives, and only Chapter 7 addresses the case of agents motivated by influence. This is because such agents can hardly be expected to provide truthful data, but rather opinions, and it is not clear that such situations actually allow accurate models.

The principle underlying all truthful mechanisms is to reward reports according to *consistency* with a *reference*. In the case of verifiable information, this reference is taken from the ground truth as it will eventually be available. In the case of unverifiable information, it will be constructed from *peer* reports provided by other agents. The trick is that the common signal that allows the agents to coordinate their strategies is the phenomenon they can all observe. Thus, by rewarding this coordination, they can be incentivized to make their best effort at observing the phenomenon and reporting it accurately. However, care must be taken to avoid other possibilities of coordination.

Another motivation for this principle was pointed out recently by Kong and Schoenebeck [4]. We can understand an agent's signal s_i and a reference signal s_j to be related only through the fact that they derive from the same underlying phenomenon. To make a prediction of s_j, the center can only use information from agent reports such as s_i. By the data processing lemma, a folk result in information theory, no amount of data processing can increase the information that signal s_i gives about s_j. In fact, any transformation of s_i other than permutations can only *decrease* the information that s_i gives about s_j. Thus, scoring the reports s_i by the amount of information they provide about s_j is a good principle for motivating agents to report information truthfully. It also aligns the agent's incentives with the center's goal of obtaining a maximum amount of information for predicting s_j.

Agent beliefs Our mechanisms use agents' self-interest to influence their strategies. This influence depends crucially on how observations influence the agent's *beliefs* about the phenomenon and about the reward it may receive for different strategies. Thus, it is crucial to model beliefs and belief updates in response to the observed signal.

The belief of an information agent a_i is characterized by a *prior* probability distribution $P_i(x) = \{p_i(x_1), \ldots, p_i(x_n)\}$ of what the state X of the phenomenon might be, where we drop the subscript if it is clear from the context.[1] Following an observation, based on the received signal s_i it will *update* its prior to a *posterior* distribution $\Pr_i(X = x|s_i = s) = P_i(x|s)$, which we often write as $Q_i(x|s)$. As a shorthand, we often drop the subscript identifying the agent and instead put the observed signal as a subscript. For example, we may write $q_s(x)$ for $q_i(x|s)$ when the agent is clear form the context. Note also that we use Pr to denote objective probabilities while P and Q are subjective agent beliefs.

Importantly, we assume that s_i is *stochastically relevant* to the states of the phenomenon $x_1, \ldots, x_n$, meaning that for all signal values $x_j \neq x_k$ $Q_i(x|x_j) \neq Q_i(x|x_k)$ so that they can be distinguished by their posterior distributions.

Belief updates We assume that agents use *Bayesian* updates where the prior belief reflects all information before observation, and the posterior belief also includes the new observation. The simplest case is when the signal s_i simply indicates a value o as the observed value. When the probability distribution is given by the relative frequencies of the different values, the update

[1]Throughout the book, we use uppercase letters to denote variables and probability distributions, and lowercase letters to denote values or probabilities of individual values.

could be weighted mean between the new observation and the prior distribution:

$$\hat{q}(x) = \begin{cases} (1-\delta)p(x) + \delta & \text{for } x = o \\ (1-\delta)p(x) & \text{for } x \neq o \end{cases} = (1-\delta)p(x) + \delta \cdot \mathbf{1}_{x=o}, \tag{1.1}$$

where δ is a parameter that can take different values according to the weight that that an agent gives to its own measurement vs. that of others.

Two properties will be particularly useful for guaranteeing properties of mechanisms that we show in this book: self-dominating and self-predicting.

Consider first objective data, where agents believe that the phenomenon they observe has one true state and they all observe this state with some measurement noise. For example, they count the number of customers in the same restaurant at a specific time, or they measure the temperature at the same place and time.

Provided the measurement is unbiased, the belief update would be to replace the prior with the probabilities obtained by the measurement, since it more accurately characterizes the actual value. This would correspond to $\delta = 1$. We could allow for the case where the agent mistrusts its observations and thus forms a convex combination of prior and observation.

As long as the measurement dominates the prior information, i.e., $\delta > 0.5$, we can show that the belief update will satisfy the following *self-dominating* condition.

Definition 1.3 An agent's belief update is *self-dominating* if and only if the observed value o has the highest probability among all possible values x:

$$q(o|o) > q(x|o) \;\; \forall x \neq o. \tag{1.2}$$

The proof is straightforward: since $\delta > 0.5$, $q(o|o) > 0.5$ and is thus larger than all other $p(x|o)$ that must be less than 0.5.

For *subjective* data, where agents observe different samples from the distribution, an agent's observation, even if absolutely certain, should not replace the prior belief, as it knows that other agents observe a different instance. For example, a customer who is dissatisfied with a product that has a very high reputation may believe that he has received a bad sample, and thus could still believe that the majority of customers are satisfied. Thus, for subjective data the observation is just one sample while the prior represents potentially many other samples.

The belief update should therefore consider an agent's observation as just one of many other samples, and give it much lower weight by using a much smaller δ. For example, $\delta = 1/t$ would compute the moving average if o is the tth observation. δ could also be chosen differently to reflect the confidence in the measurement. We call this update model the *subjective* update.

Clearly, subjective belief updates do not always satisfy the self-dominating condition, as is already shown in the example of poor service. Thus, we introduce a weaker condition by only requiring the *increase* in the probability of the observed value to be highest.

Definition 1.4 An agent's belief update is *self-predicting* if and only if the observed value has the highest relative increase in probability among all possible values:

$$q(o|o)/p(o) > q(x|o)/p(x) \;\; \forall x \neq o. \tag{1.3}$$

This condition is satisfied when an agent updates its beliefs in a subjective way as in Equation (1.1), since its reported value o is the only one that shows in increase over the prior probability.

In a more general scenario, an agent may obtain from its observation a vector $p(obs|x)$ that gives the probability of its observation given the different possible values x of the phenomenon. Using the maximum likelihood principle, it would report the value o that assigns the highest probability to the observation:

$$o = \underset{x}{\text{argmax}}\; p(s_i = obs|x).$$

For its belief update, Bayes rule also allows to compute a vector u of probabilities for each of the values:

$$u_i(x) = p(x|s_i = obs) = \frac{p(s_i = obs|x)p(x)}{p(s_i = obs)}$$

where $p(s_i = obs)$ is unknown but can be obtained from the condition that $\sum_x u_i(x) = 1$. A Bayesian agent will use this vector for its belief update:

$$\hat{q}_i(x) = (1-\delta)p(x) + \delta u_i(x) = (1 - \delta + \delta \alpha p(s_i = obs|x))p(x), \tag{1.4}$$

where $\alpha = 1/(\sum_x p(s_i = obs|x)p(x))$ is a normalizing constant. This form of updating has the capability of taking into account correlated values. For example, when measuring a temperature of 20°, due to measurement inaccuracies 19 and 21 may also increase their likelihood. This could be reflected in increases in the posterior probability not only for 20, but also the neighboring values 19 and 21.

Provided that the agent chooses the value o it reports according to the maximum likelihood principle, this more precise update also satisfies the self-predicting condition, since $\hat{q}_i(x)/p(x) = (1 - \delta + \delta\alpha p(s_i = obs|x))$ is largest for the maximum likelihood estimate o that maximizes $p(s_i = obs|x)$. However, even for high δ there is no guarantee of satisfying the self-dominating condition, since the prior probability of the maximum-likelihood value could be very small.

Deriving agent beliefs from ground truth One way to model an agent's observation of objective data is that it observes the ground truth through a *filter* that has certain noise and systematic biases. Sometimes there exists a model of this filter in the form of a *confusion matrix* $F(s|\omega) = \Pr(s_i = s|\Omega = \omega)$ that gives the probability of an observed signal s_i given a ground truth ω, shown as an example in Figure 1.8. Such models have proven useful to correct noise in

		Phenomenon state ω		
		a	b	c
agent	a	0.8	0.2	0
signal	b	0.2	0.6	0.2
s	c	0	0.2	0.8

Figure 1.8: Example confusion matrix, giving the probability distribution $F(s|\omega) = Pr(s_i = s|\Omega = \omega)$.

postprocessing data; for example, Dawid and Skene [5] show a method that uses the expectation maximization algorithm to construct an optimal estimation of underlying values that is consistent with the signal reports. Such a model could let us derive what an agent's beliefs and belief updates *should* be. This confusion matrix could arise, for example in the following crowdworking scenario: agents are asked to classify the content of web pages into inoffensive (value a), mildly offensive (b), and strongly offensive (c). The prior distribution for this example might be:

$P(\omega)$

p(a)	p(b)	p(c)
0.79	0.20	0.01

and the confusion matrix of Figure 1.8 might characterize the observation bias applied by crowdworkers, who tend to err on the side of caution and classify inoffensive content as offensive. Note that given this observation bias, the prior distribution of the signals shows this bias:

$P(s_i)$

p(a)	p(b)	p(c)
0.67	0.28	0.05

For our purposes, knowing the confusion matrix of the agents reporting the information allows computing the posterior beliefs that an agent *should have* about its peers following an observation. Using the confusion matrix and the prior probabilities given above:

$$\begin{aligned} q(s_j|s_i) &= \sum_{\omega\in\Omega} f(s_j|\omega) f(\omega|s_i) \\ &= \sum_{\omega\in\Omega} f(s_j|\omega) f(s_i|\omega) \frac{p(\omega)}{p(s_i)} \\ &= \frac{\sum_{\omega\in\Omega} f(s_j|\omega) f(s_i|\omega) p(\omega)}{\sum_{\omega\in\Omega} f(s_i|\omega) p(\omega)}, \end{aligned}$$

where f refers to the probabilities defined by the confusion matrix model in Figure 1.8. Figure 1.9 gives the posterior probabilities $q(s_j|s_i)$ assuming the shown prior probabilities for each value. This prediction can be useful for designing incentive mechanisms, or for understanding

		Agent i's observation s_i		
		a	b	c
agent j's	a	0.77	0.54	0.17
observation	b	0.22	0.37	0.53
s_j	c	0.01	0.09	0.3

Figure 1.9: Resulting belief updates: assuming the shown prior beliefs and the confusion matrix shown in Figure 1.8, an agent would form the posterior probabilities $q(s_j|s_i)$ shown.

conditions on agent beliefs that are necessary for such mechanisms to work.

We can see that for predicting the observation of the peer agent j, value a is dominating for observations a and b, and b is dominating for observation c. Thus, the distributions are clearly not self-dominating. Are they at least self-predicting? Using the same confusion matrix and prior probabilities as above, we can obtain the relative increases of probability $q(s_j|s_i)/p(s_j)$ as the following matrix:

		Agent i's observation s_i		
		a	b	c
agent j's	a	1.137	0.80	0.25
observation	b	0.80	1.32	1.90
s_j	c	0.25	1.90	6.25

Clearly, the values on the diagonal are the highest in their respective columns, meaning the the observed value also sees that largest increase in probability for the peer observation. Thus, the distribution satisfies the self-predicting property.

However, this is not always the case. When the proportion of errors increases further, even the self-predicting condition can be violated. Consider, for example, the following confusion matrix:

		World state Ω		
		a	b	c
agent	a	0.8	0.2	0
observation	b	0.1	0.5	0.3
s_i	c	0.1	0.3	0.7

This leads to the following prior probabilities of the different observations:

$P(s_i)$		
p(a)	p(b)	p(c)
0.67	0.18	0.15

and so we can obtain the relative increases of probability $p(s_j|s_i)/p(s_j)$ as the following matrix:

		Agent i's observation s_i		
		a	b	c
agent j's	a	1.137	0.68	0.77
observation	b	0.68	1.78	1.51
s_j	c	0.77	1.50	1.44

and it turns out that this belief structure is not self-predicting since $p(b|c)/p(c) = 1.51 > p(c|c)/p(c) = 1.44$. This increased likelihood of b happens because mildly offensive content is often misclassified as offensive. As mildly offensive content is much more likely than offensive content, it is the most likely cause of an "offensive" signal, and the most likely peer signal.

What theoretical guarantees can be given for belief updates given based on modeling filters in this way? We present three simple cases, the first valid for general situations and the other two for binary answer spaces. For the self-dominating condition, we can observe the following.

Proposition 1.5 *Whenever for all agents i and all x, $f(s_i = x|\Omega = x)$ and $f(\Omega = x|s_i = x)$ are both greater than $\sqrt{0.5} = 0.71$, then the belief structure satisfies the self-dominating condition, even if agents have different confusion matrices and priors.*

Proof. The conditional probability $q(s_j|s_i)$ for $s_i = s_j = x$ is:

$$\begin{aligned} q(s_j = x|s_i = x) &= \sum_w f(s_j = x|\Omega = w) f(\Omega = w|s_i = x) \\ &> f(s_j = x|\Omega = x) f(\Omega = x|s_j = x) \\ &\geq 0.5. \end{aligned}$$

Since $q(s_j = x'|s_i = x) \leq 1 - q(s_j = x|s_i = x) < 0.5$, we obtain that $q(s_j = x|s_i = x) > q(s_j = x'|s_i = x)$. □

To ensure the self-predicting condition, we can impose a weaker condition, provided that agents have identical confusion matrices and priors.

Proposition 1.6 *For binary answer spaces and identical confusion matrices and priors, whenever $F(s|\omega)$ is fully mixed and non-uniform, belief updates satisfy the self-predicting condition.*

For conditional probability $q(s_j|s_i)$ and $s_i = s_j = x$ we have:

$$q(s_i = x|s_j = x) = \sum_{\omega} f(s_i = x|\Omega = \omega) \cdot f(\Omega = \omega|s_j = x) = \sum_{\omega} f(x|\omega)^2 \cdot \frac{p(\omega)}{p(x)}$$
$$= \frac{\sum_{\omega} f(x|\omega)^2 \cdot p(\omega)}{\sum_{\omega} f(x|\omega) \cdot p(\omega)} > \frac{\left(\sum_{\omega} f(x|\omega) \cdot p(\omega)\right)^2}{\sum_{\omega} f(x|\omega) \cdot p(\omega)},$$

where the inequality is due to Jensen's inequality, with the strictness following from the condition that $F(s|\omega)$ is fully mixed and non-uniform.

Therefore:

$$q(s_i = x|s_j = x) > \sum_{\omega} f(x|\omega) \cdot p(\omega) = p(s_i = x) = 1 - p(s_i = y)$$
$$> 1 - q(s_i = y|s_j = y) = q(s_i = x|s_j = y)$$

where the last inequality is due to $q(s_i = y|s_j = y) > p(s_i = y)$.

For heterogeneous beliefs, we can ensure the self-predicting condition under a slightly stronger condition, but only valid for binary answer spaces.

Proposition 1.7 *For binary answer spaces and heterogenous confusion matrices and priors, whenever $p(s = x|\Omega = x) > p(s = x)$, belief updates satisfy the self-predicting condition.*

Proof. Notice that:

$$q(s_i = x|s_j = z) = f(s_i = x|\Omega = x) \cdot f(\Omega = x|s_j = z) + f(s_i = x|\Omega = y) \cdot f(\omega = y|s_j = z)$$
$$= [f(s_i = x|\Omega = x) - f(s_i = x|\Omega = y)] \cdot f(\Omega = x|s_i = z) + f(s_i = x|\Omega = y).$$

Since:

$$f(s_i = x|\Omega = x) - f(s_i = x|\Omega = y) = f(s_i = x|\Omega = x) - 1 + f(s_i = y|\Omega = y)$$
$$> f(s_i = x) - 1 + f(s_i = y) = 0$$

z that (strictly) maximizes $q(s_i = x|s_j = z)$ is equal to z that (strictly) maximizes $f(\Omega = x|s_j = z)$. Using Bayes rule and the condition of the proposition, we obtain:

$$f(\Omega = x|s_j = x) = \frac{f(s_j = x|\Omega = x)}{p(s_j = x)} \cdot p(\Omega = x) > \frac{p(s_j = x)}{p(s_j = x)} \cdot p(\Omega = x) = p(\Omega = x)$$
$$f(\Omega = x|s_j = y) = \frac{f(s_j = y|\Omega = x)}{p(s_j = y)} \cdot p(\Omega = x) < \frac{p(s_j = y)}{p(s_j = y)} \cdot p(\Omega = x) = p(\Omega = x)$$

so it must also hold that $f(\Omega = x|s_j = x) > f(\Omega = x|s_j = y)$. Therefore, $z = x$ strictly maximizes $p_j(\Omega = x|z)$ and consequently $p(x|z)$. □

NOTATION

Notation	Meaning
$P, Q, R, \ldots$	uppercase: probability distribution
$p(x), q(x), r(x), \ldots$	lowercase: probability of value x
$E_P[f(x)]$	expected value of $f(x)$ under distribution P: $\sum_x p(x) \cdot f(x)$
$H(P)$	entropy of probability distribution P, $H(P) = \sum_x -p(x) \log p(x)$
$D_{KL}(P\|\|Q)$	Kullback-Leibler Divergence $D_{KL}(P\|\|Q) = \sum_x p(x) \log \frac{p(x)}{q(x)}$
$\lambda(P)$	Simpson's diversity index $\lambda(P) = \sum_x p(x)^2$
$\mathbf{1}_{cond}$	selector function: $1_{cond} = 1$ if *cond* is true, $= 0$ otherwise
$f(-x)$, f_{-x}	f is a function that is independent of x
$freq(x)$	frequence of value x normalized so that $\sum_x freq(x) = 1$
$gm(x_1, \ldots, x_n)$	geometric mean of $x_1, \ldots, x_n$, $gm(x_1, \ldots, x_n) = \sqrt[n]{x_1 \cdot \ldots \cdot x_n}$

ROADMAP

This books presents an overview of incentives for independent self-interested agents to accurately gather and truthfully report data. We do not aim to be complete, but our main goal is to make the reader understand the principles for constructing such incentives, and how they could play out in practice.

We distinguish scenarios with *verifiable* information, where the mechanism learns, always or sometimes, a ground truth that can be used to verify data, and *unverifiable* information, where a ground truth is never known.

When information is verifiable, incentives can be provided to each agent individually based on the ground truth, and we describe schemes in Chapter 2. When information is not verifiable, incentives can still be derived from comparison with other agents through a game-theoretic mechanism. However, this necessarily involves assumptions about agent beliefs. Thus, in Chapter 3, we describe mechanisms where these assumptions are parameters of the mechanism that have to be correctly set from the outside. Often, setting these parameters is problematic, so there are *nonparametric* mechanisms that obtain the parameters either from additional data provided by the agents, or from statistics of the data provided by a group of agents. We present mechanisms that follow these approaches in Chapters 4 and 5.

As verification also allows assessing the influence of data on the output of a learning algorithm, incentives can be used to align incentives of agents with those of the learning mechanism. One way is to reward agents for their positive influence on the model through prediction markets, a technique we describe in Chapter 6. Another is to limit their influence on the learning outcome so as to thwart malicious data providers whose goal is to falsify the learning outcome.

We discuss how maintaining reputation can achieve this effect in Chapter 7. In Chapter 8, we consider issues that present themselves when the techniques are integrated into a machine learning system: managing the information agents and self-selection, scaling payments and reducing their volatility, and integration with machine learning algorithms.

CHAPTER 2

Mechanisms for Verifiable Information

The simplest case is when the accuracy of data can be verified later, so that rewards can be given based on this information once it becomes available. This case is actually quite frequent. It includes forecasts of weather, product sales, election outcomes and many other phenomena. Sometimes environmental measurements, such as pollution, can be verified on a cumulative basis at low cost. Crop diseases eventually become apparent.

In this chapter, we are going to consider two types of incentive mechanisms: eliciting a single value, and eliciting a probabilty distribution of values.

2.1 ELICITING A VALUE

We consider that the center would like to know or predict the value of a phenomenon $X \in \{x_1, \ldots, x_N\}$, that we limit to *discrete* values. The scenario is as shown in Figure 1.7; agents can observe the phenomenon, obscured by measurement noise. The center will obtain the ground truth $g \in \{x_1, \ldots, x_N\}$ at a later time and can use it to compute the reward.

Here are some examples that fit this model.

- *Is the crop at location l diseased?* Possible answers are x_1 for no disease, and $x_2 = disease_1, \ldots, x_N$ for the different known diseases.
- *Who will win the presidential election?* Possible answers are $x_1 = cand_1, \ldots, x_N = cand_n$ for the different candidates.
- *How much will this product sell?* Possible answers are $x_i, i \in \{1, \ldots, N\}$ where x_i means the product sales are between $(i-1)$ and i million.

A basic mechanism that incentivizes cooperative strategies in this example, is the basic *truth agreement mechanism*, shown as Mechanism 2.1.

We claim that this mechanism will cause rational agents to adopt a cooperative reporting strategy. To understand why this is the case, we have to consider how it makes the setting look from an agent's perspective, as illustrated in Figure 2.1.

1. The agent has a *prior* probability distribution $p_i(x)$, with $x_0 = \text{argmax}\, p_i(x)$ the most likely value. We abbreviate $p_i(x) = p(x)$.

Mechanism 2.1 Truth agreement.

1. Agent reports data = (discrete) value x.
2. Center observes the ground truth g (at some later time).
3. Center pays agent a reward:

$$pay(x, g) = \begin{cases} 1 & \text{if } x = g \\ 0 & \text{otherwise.} \end{cases}$$

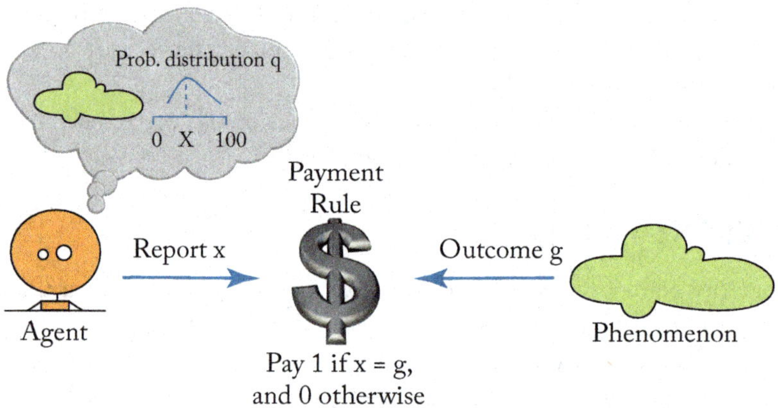

Figure 2.1: Truth agreement mechanism seen from an agent's perspective.

2. The agent i observes a signal s_i and obtains a *posterior* distribution $q_i(x|s_i)$. We abbreviate $q_i(x|s_i) = q(x|s)$ or $q(x)$. $x_1 = \text{argmax}\, q(x)$ is now the most likely value. Therefore, the agent believes that x_1 has highest probability $q(x_1)$ to match g.

3. The agent reports x_1 and expects (future) payoff $q(x_1)$.

Two issues complicate this analysis: agents that do not observe the phenomenon can also expect a positive payoff, and the fact that observations have a cost.

Discouraging heuristic reports With the basic truth agreement mechanism, an agent that does not observe the phenomenon could use a heuristic strategy: simply report the value x_0 that is most likely according to its prior, and still expect a positive reward $p(x_0)$. This would mean that the center could be swamped by such uninformed reports! It is important to reduce the payment so that heuristic strategies have an expected reward of 0. Note that this is different from concerns about the budget that the center has to spend to pay for the data, which has to be dealt with through scaling.

We can do this by modifying the basic truth agreement mechanism to subtract the expected reward when reporting according to the prior, $E_{prior}[pay] = p(x_0)$, for example, by charging a fee for an agent to participate in the mechanism. Estimating $p(x_0)$ is often not easy, although there are some background constraints, for example, with N values we know that $p(x_0) \geq 1/N$ as at least one value must have higher than average probability. In Chapter 5, we will see the correlated agreement mechanism that automatically determines this value from agents' reports.

Costly measurements Assume that the agent incurs a cost m for observing the signal s_i about the phenomenon. If the cost is very high, or if the signal does not carry much information, there is a danger that it might skip the observation. To avoid this problem, we have to ensure that m does not exceed

$$\underbrace{q(x_1)}_{E_{post}[pay]} - \underbrace{p(x_0)}_{E_{prior}[pay]} .$$

To achieve this, we need to scale the payment by $\alpha \geq \frac{m}{q(x_1)-p(x_0)}$. Note that this α depends on the measurement technology and agent's confidence, and may be complex to determine. However, it turns out that in general it will become apparent in agents' behavior: if α is too low, rational agents will not participate as their expected reward will not be greater than zero. Thus, in general, we can tune α by gradually increasing it until we get sufficient participation.

Components of payment schemes To summarize, the final payment rule of the truth agreement mechanism is as follows:

$$\begin{aligned} pay(x,g) &= \alpha \left[-p(x_0) + \begin{cases} 1 & \text{if } x = g \\ 0 & \text{otherwise} \end{cases} \right] \\ &= \alpha \left[\mathbf{1}_{x=g} - p(x_0) \right] \end{aligned}$$

with the following components:

- a reward for delivering truthful data: pay 1 if the ground truth is matched. Note the notation $\mathbf{1}_{x=g}$ that we shall adopt in the rest of this book to denote such agreement-based functions;
- an offset to make the expected reward of heuristic reports equal to zero; and
- a scaling factor α to compensate for the cost of measurement.

The agent knows about the phenomenon only through the signal it observes, and so in its expectation it delivers truthful data by reporting the value that is the most likely according to the posterior distribution q it derives from this signal. To also achieve truthfulness from the center's perspective, both the agent and the center must interpret the phenomenon in the same way. For example, when reporting temperature there must be agreement on where and when it is measured, and what scale (Celsius, Fahrenheit, Kelvin) is used.

Throughout the book, we will mainly focus on the first component: reporting truthful data. The offset to discourage heuristic reports may have to be estimated by the center; many of the mechanisms we present will include this component in their design.

When observing the phenomenon to obtain truthful data involves cost, the payment will have to be scaled so that the incentive exceeds this cost. Note that this scaling is always possible as long as the incentive is strict: the difference in reward between truthful and non-truthful report is strictly positive. Thus, *throughout this book we will focus on ensuring strict truthfulness*, as the scaling will depend very much on the application and technology used by information agents. It is also possible to sharpen incentives by combining rewards for multiple data items; we shall discuss such possibilities in Chapter 8.

We formally state the properties of the Mechanism 2.1 with the scaling given above as follows.

Theorem 2.1 *Provided the scaling factor α is large enough, the scaled Truth Agreement Mechanism induces dominant strategies that are cooperative. With the correct offset, heuristic strategies carry no expected revenue.*

In the theorem, the notion of dominant strategy means a solution concept in which an agent is incentivized to adopt the specified reporting strategy regardless of the strategies of other agents. We note that similar mechanisms can be designed for eliciting not only observed values, but also properties of multiple values such as their mean or mode; see [6, 7] for more details.

Expected payment From the perspective of the information agent, the payment it can expect for a measurement x is $\alpha[\max_y(q(y|x)) - p(x_0))]$. As $p(x_0)$ is constant and fixed by the mechanism, the payment varies as $\max_x q(x)$.

Truth agreement in crowdsourcing: Gold tasks A common way of discouraging lazy or uncooperative workers in crowdsourcing is to mix within the tasks some *gold* tasks. These are tasks that agents cannot distinguish from others, but where the requester knows the answer and can use this to check on the performance of the worker.

The simplest way to use gold tasks is by using truth agreement: workers can collect bonuses for answering gold tasks correctly. Since they do not know whether a task is a gold task, this translates into a (weaker) incentive for all tasks. However, for the incentives to be strong enough to motivate workers to exert effort, either there must be a significant number of gold tasks, or the bonus associated with each gold task must be very high. Having a large number of gold tasks wastes worker effort, while having a high bonus is problematic because it increases the volatility of the bonus.

de Alfaro et al. [8] propose a solution where only a few gold tasks are used to incentivize a highest layer of workers. Since one can assume that they will put in the required effort to provide good answers throughout, their answers can then be used as gold tasks for a next layer of workers, who can in turn provide the gold tasks for the next level, and so forth. In this way, a small number of gold tasks can incentivize a whole hierarchy of workers. One issue with the method is that workers higher in the hierarchy must be more proficient, which may be hard for the requester to determine.

2.2 ELICITING DISTRIBUTIONS: PROPER SCORING RULES

Mechanism 2.2 Scoring rule.

1. Agent reports data = probability distribution A.
2. Center observes the ground truth g (at some later time).
3. Center pays agent a reward:

$$pay(A, g) = SR(A, g),$$

where SR is a *proper scoring rule.*

What if we would like the agents to not just report the value they consider most likely, but their posterior probability distribution q (Figure 2.1)? For example, we would like a weather forecaster to not just tell us what the weather next Sunday will most likely be, but also give us an indication of confidence by providing us with a complete probability distribution.

Suppose that the prior probability, as obtained from historical averages, is:

$$p = \frac{\begin{array}{c|c|c} Rain & Cloud & Sun \end{array}}{\begin{array}{c|c|c} 0.2 & 0.3 & 0.5 \end{array}}$$

Now our agent, with access to meterological data, examines this data and forms the posterior distribution:

$$q = \frac{\begin{array}{c|c|c} Rain & Cloud & Sun \end{array}}{\begin{array}{c|c|c} 0.8 & 0.15 & 0.05 \end{array}}$$

Using our truth agreement mechanism, we could get the agent to tell us that rain is most likely. But since we were really hoping to go on a picnic, we are not happy with such an answer and want to take our chances to maybe enjoy a sunny day after all. How do we get the agent to also tell us its probability estimate for sunny weather?

The answer is to use Mechanism 2.2. It is based on *proper scoring rules*, first proposed by Brier [9] and Good [10], and illustrated in Figure 2.2. A proper scoring rule scores a reported

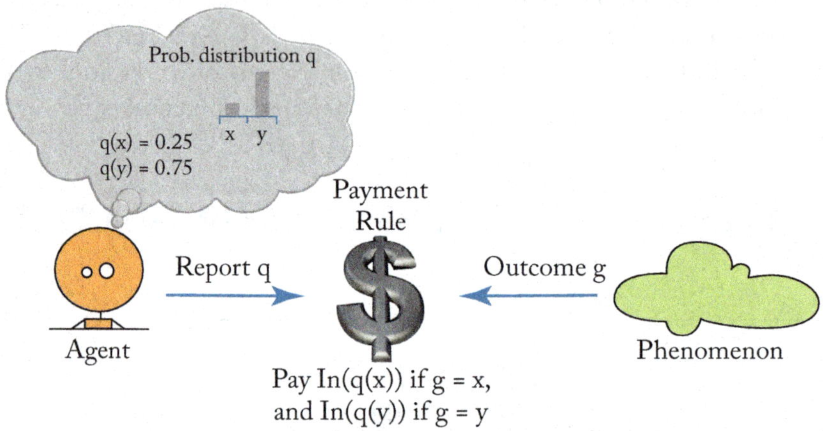

Figure 2.2: Mechanism using a proper scoring rule.

probability distribution A against a ground truth g and provides a payment such that

$$(\forall \underline{q'} \neq \underline{q}) \sum_x q(x) \cdot pay(\underline{q}, x) > \sum_x q(x) \cdot pay(\underline{q'}, x).$$

Well-known examples are

- the quadratic scoring rule [9]:

$$pay(\underline{A}, g) = 2 \cdot A(g) - \sum_{x \in X} A(x)^2; \quad \text{and}$$

- the logarithmic scoring rule [10]:

$$pay(\underline{A}, g) = C + \ln A(g).$$

An overview of proper scoring rules and their properties can be found in Geiting and Raftery [11].

In our weather example, suppose that the agent provides us with its true distribution. Depending on the weather we actually observe on Sunday, the agent would receive the following payments.

Sunday's weather	Payoff (log)	Payoff (quadratic)
Rain	$C + \ln 0.8 = C - 0.22$	$2 \cdot 0.8 - 0.665 = 0.935$
Cloud	$C + \ln 0.15 = C - 1.89$	$2 \cdot 0.15 - 0.665 = 0.935 = -0.365$
Sun	$C + \ln 0.05 = C - 3.0$	$2 \cdot 0.05 - 0.665 = -0.565$
Average	$C - 0.6095$	0.665

where for the quadratic scoring rule we used the fact that $(0.8^2 + 0.15^2 + 0.05^2) = 0.665$.

What offset do we need to choose to ensure that the expected reward for uninformed reporting is equal to zero? Note that when using the logarithmic scoring rule, the expected reward for reporting the prior distribution P is equal to:

$$E[Pay(P)] = \sum_x p(x)(C - \ln p(x)) = C - H(P),$$

where $H(P)$ is the entropy of the prior distribution P. Thus, we should set the constant to $C = H(P)$. The expected reward for an agent is thus $\alpha[H(P) - H(Q)]$ where α a constant chosen to compensate effort. Note that the reward is thus proportional to the information that the agent's measurement gives about the true signal.

For the quadratic scoring rule, the expected payment for reporting according to the prior is

$$E[Pay(P)] = \sum_x p(x)(2p(x) - \sum_y p(y)^2) = \sum_x p(x)^2$$

which is equal to Simpson's diversity index [12] $\lambda(P)$. Thus, we should subtract $\lambda(P)$ and the expected reward is $\alpha[\lambda(Q) - \lambda(P)]$.

Why do proper scoring rules motivate cooperative reporting strategies? Let's consider the expected reward using the logarithmic scoring rule:

$$E[pay(\underline{A}, g)] = \sum_x q(x) \cdot pay(\underline{A}, x) = \sum_x q(x) \cdot [C + \ln(a(x))]$$

and the difference between truthful and non-truthful reports:

$$\begin{aligned} E[(pay(\underline{Q}, g)] & -E[pay(\underline{Q}', g)] \\ = & \sum_x q(x) \cdot [C + \ln q(x)) - (C + \ln q'(x))] \\ = & \sum_x q(x) \cdot \ln \frac{q(x)}{q'(x)} \\ = & D_{KL}(\underline{Q} || \underline{Q}'). \end{aligned}$$

By Gibbs' inequality, $D_{KL}(Q||Q') \geq 0$ with equality only when $Q = Q'$. Reporting a $\underline{Q}' \neq \underline{Q}$ can therefore only get a lower payoff than truthfully reporting Q.

Similarly, for the quadratic scoring rule, we have:

$$E[pay(\underline{A}, g)] = \sum_x q(x) \cdot pay(\underline{A}, x) = \sum_x q(x) \cdot [2a(x) - \sum_z a(z)^2]$$

and the difference between truthful and non-truthful reports:

$$
\begin{aligned}
E[(pay(\underline{Q}, g)] \quad & -E[pay(\underline{Q}', g)] \\
= \quad & \sum_x q(x) \cdot [2(q(x) - q'(x)] - \sum_z [q(z)^2 - q'(z)^2] \\
= \quad & \sum_x [q(x)^2 - 2q(x)q'(x) + q'(x)^2] \\
= \quad & \sum_x [q(x) - q'(x)]^2
\end{aligned}
$$

which again is greater than 0 except with $\underline{Q} = \underline{Q}'$, so the expected reward is maximized when truthfully reporting Q.

We summarize these results about Mechanism 2.2 with proper scaling as follows:

Theorem 2.2 *For both the logarithmic and the quadratic scoring rule, with proper scaling the scoring rule mechanism induces dominant reporting strategies that are cooperative. With the proper offset, the expected payoff for heuristic reporting is equal to zero.*

CHAPTER 3

Parametric Mechanisms for Unverifiable Information

In most cases, the ground truth is never known. For hotels and restaurants, there is no neutral evaluator that can verify the correctness of reviews. In distributed sensing, many quantities are never measured other than through crowd sensors. For predictions about hypothetical questions, we never obtain the ground truth.

There is another complexity that is introduced when we cannot verify the ground truth, illustrated by Figure 3.1. When information can be verified, such as a temperature measurement, it is always *objective*: all agents observe exactly the same variable, which corresponds to the ground truth. When it is unverifiable, it may also be *subjective*: agents observe different *samples*, all drawn from the same distribution. This is the case for example for restaurant reviews: every customer gets a different meal, but cooked by the same chef. In such a situation, there is no ground truth for individual data items. However, there is a ground truth for the *distribution* of these items.

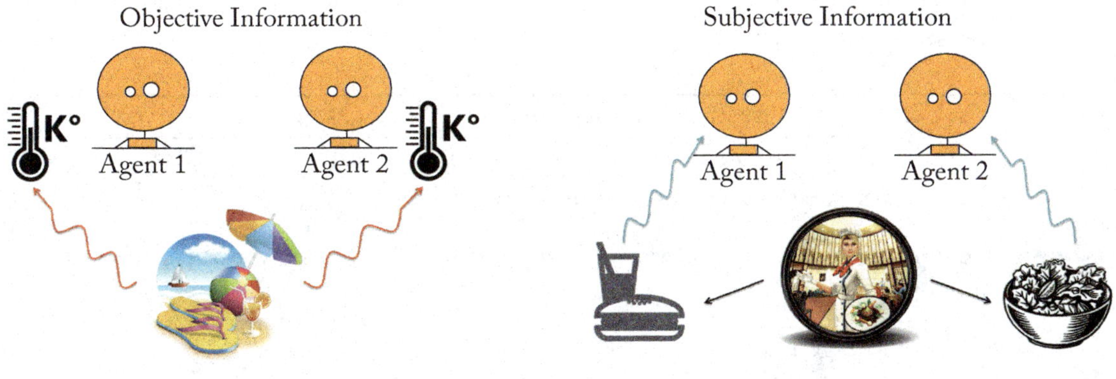

Figure 3.1: Objective vs. subjective information.

As a consequence of this distinction, the goals of the elicitation process also vary: while for objective information, the goal is to obtain the particular *value* as accurately as possible, for subjective information the goal is to obtain an accurate *distribution*.

Peer consistency mechanisms To validate data in such scenarios, we need to use its coherence with data submitted by other agents that observe the same phenomenon. We call such agents *peer* agents and the class of incentive schemes based on peer reports *peer consistency*. Figure 3.2 illustrates the principle. We first consider peer consistency mechanisms for the simpler case of objective information.

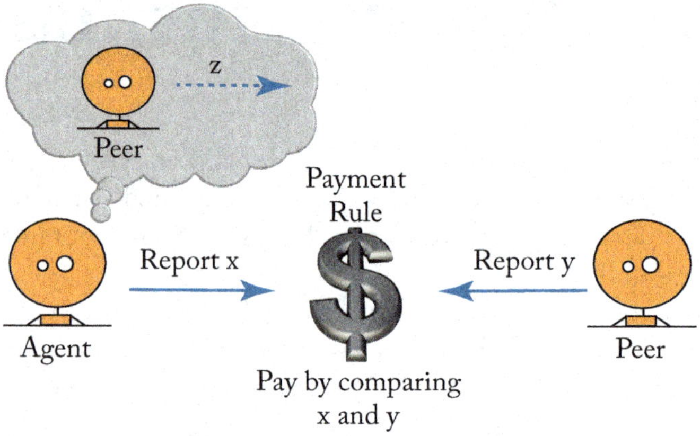

Figure 3.2: Peer consistency scenario.

3.1 PEER CONSISTENCY FOR OBJECTIVE INFORMATION

3.1.1 OUTPUT AGREEMENT

Mechanism 3.1 The output agreement mechanism.

1. Center gives a task to agents a_i; a_i reports data x_i.
2. Center randomly selects a *peer agent* a_j that has also been given the same task and reported data x_j.
3. Center pays agent a_i a reward:

$$pay(x_i, x_j) = \begin{cases} 1 & \text{if } x_i = x_j \\ 0 & \text{otherwise} \end{cases} .$$

A well-known peer consistency mechanism that has been widely used to elicit objective information from multiple agents is *output agrement*, a term coined in von Ahn and Dabbish

[13]. It is shown in Mechanism 3.1. It involves giving the same task to two agents, and paying them a constant reward if and only if they give the same answer. This mechanism was popularized in the ESP game for labeling images with keywords, shown in Figure 3.3.

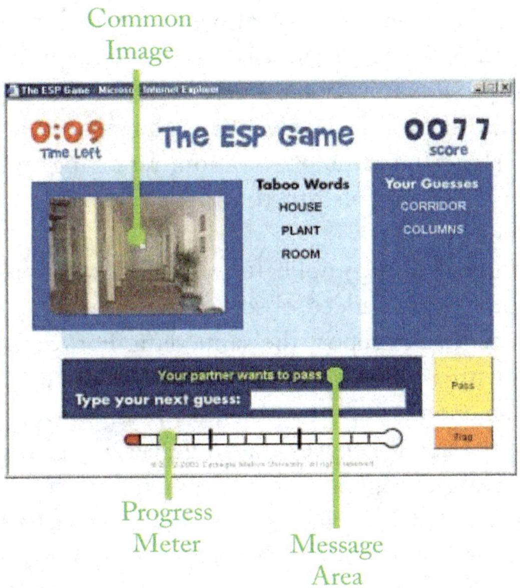

Figure 3.3: Screenshot of a stage in the ESP game.

In this game, an image is shown to a person who has to input keywords that describe what is seen in the image. Two people observe the same image at the same time, and the keywords they type are compared to one another. Only if the keywords match, they obtain points as a reward. Certain words are declared *taboo* to avoid coordination on trivial words that do not provide much information.

In peer consistency mechanisms such as output agreement, reporting data becomes a *game* among agents, as the reward depends both on the actions of the agent submitting the data *and* on those of its peer agents. Where before we could consider each agent's choice of strategy as an optimization by itself, now the optimal strategies chosen by different agents depend on each other.

Note that in an actual application, we may not want to give the same task to multiple agents, since it increases the cost of information gathering. The peer report is thus often obtained from peer reports on related tasks that are translated through a model. For example, for a pollution measurement the peer report might be constructed by an interpolation between several reports received for locations in the vicinity. However, analyzing the properties of such interpolation is too complex, and we therefore generally assume that there is one particular peer agent.

3.1.2 GAME-THEORETIC ANALYSIS

Such situations are analyzed by game theory, and it is generally considered that the chosen strategies should form an *equilibrium*: no agent can improve its expected payoff by deviating from the equilibrium strategies, provided no other agent deviates. For example, if we use a reward mechanism where we reward data if and only if it is the same as that reported by a random peer agent, and all agents share the same observation, reporting this observation truthfully is an equilibrium: if the peer agents are also truthful, our data will match and thus reporting it accurately is the best strategy.

However, things become more complicated when observations are noisy: now an agent cannot be so sure that a peer agent indeed observes the same signal, and so even if it makes an effort to be truthful, it might not report the same data. For example, if we are measuring a temperature, and report it with a precision of 1 degrees, it is quite likely that a peer will obtain a slightly different measurement and report a non-matching value.

However, as long as the agent believes that the peer agent measures in the same way as itself, the value that a truthful peer agent is most likely to report is the same as that observed by the agent. Thus, we have a *Bayesian* game where there is uncertainty about the value of the phenomenon, and agents share a belief about the distribution of this value. The game will have at least one *Bayes-Nash equilibrium*, and as shown by the argument above cooperative reporting is one such equilibrium.

More formally, as defined in the introduction, let $p_i(x)$ be agent i's prior belief about the phenomenon x, and $q_i(x)$ its posterior belief after observation. When observing objective data in an unbiased way, we may assume that the agent's beliefs are *self-dominating* as in Definition 1.3. That is, the agent believes that her peer is most likely to observe the same value she does.

Note that this condition makes no assumption about the absolute strengths of agent beliefs, but only about the relative strengths, and it allows the posterior distributions to be quite different.

The game-theoretic notion of Bayes-Nash equilibria assumes that all agents have a *common* prior belief, and would require all their posterior beliefs to be identical as well. However, we can see that the output agreement mechanism does not require such a strict condition. Thus, we use an extended notion called *ex-post subjective Bayes-Nash equilibrium*, introduced in Witkowski and Parkes [19].

Definition 3.1 A set of strategies $\underline{s}$ is an *ex-post subjective Bayes-Nash equilibrium* (PSBNE) if it forms a Bayes-Nash equilibrium for all combinations of admissible agent beliefs.

For example, we might define an admissible agent belief as one where it has a common prior belief p, and the update to posterior beliefs satisfies the self-dominance condition described above. We can then prove:

Theorem 3.2 *For self-dominating belief updates, the output agreement mechanism shown in Mechanism 3.1 has a strict ex-post subjective Bayes-Nash equilibrium in cooperative strategies.*

The proof is straightforward: provided that its peer agent j adopts cooperative strategy, an agent i, who observes x, believes that the most likely value that j observes (and reports) is equal to x due to the self-dominant condition. Therefore, agent i will most likely get a reward if it also adopts a cooperative strategy and reports x.

Let us appreciate this result: we now have a mechanism that will ensure, in a wide range of conditions, that rational agents will make the effort to report objective data as accurately as possible, *even if we can never verify it!* The trick is that we involve the agents in a game where the winning strategy is to coordinate the data they report, and they all need to accurately measure the phenomenon to achieve this coordination. This shows the great potential offered by involving multiple agents, but unfortunately there are some complications, as we shall see below.

When agents' observations are not perfect, but subject to a minimum error probability that depends on the effort they exert, the center has to scale the rewards sufficiently so that agents exert their best effort. Liu and Chen [14] show, for tasks with binary answers, how to determine the minimal reward level for the output agreement mechanism such that exerting maximal effort is an equilibrium. The analysis is complex and involves many assumptions so we do not discuss it in detail here.

Uninformative equilibria Games often have multiple equilibria. What we have shown so far is that one of the equilibria in the game induced by the output agreement mechanism is to adopt cooperative strategies. However, there are also other equilibria where agents adopt heuristic strategies. In particular, there are equilibria where every agent always reports the same identical value, regardless of what it observed. Even worse, such equilibria have higher payoffs: there is no measurement uncertainty, and often no cost of measurement either [16].

We call equilibria in heuristic strategies *uninformative* equilibria, since the agents provide no information to the center when they adopt them.

There are several ways to avoid such equilibria. In the ESP game, certain common words that were obvious candidates for such equilibria are considered "taboo" and do not result in any points. In general, one could penalize agents if all reports are uniform, and thus eliminate such equilibria. However, there are more elegant solutions when mechanisms can use multiple peer reports, as we will see later in this book.

Parametric mechanisms The output agreement is a non-parametric mechanisms, meaning that it does not contain any parameters related to agents' beliefs. Unfortunately, such a restriction, does allow existence of a general truthful mechanism, as formally expressed by the following theorem.

Theorem 3.3 *There does not exist a* non-parametric mechanism *that has cooperative reporting as a strict Bayes-Nash equilibrium for a general structure of agents' beliefs.*

The result was first shown in the context of uncommon beliefs [20], and was later shown to hold even if agents have a common belief [34, 36]. This shows the limitations of nonparametric mechanisms: even under a fairly constrained set of possible beliefs, it is not possible to properly elicit private information, unless the mechanism has some knowledge about agents' beliefs, encoded in its parameters.

The proof of the theorem is out of the scope of this book, but its intuition is fairly simple. If a non-parametric mechanism elicits honest responses for one set of agents' beliefs, then there exists another set of agents' beliefs for which the mechanism fails to provide proper incentives. Namely, agents' expected payoff crucially depends on their beliefs, so unless a mechanisms has at least partial knowledge of agents' beliefs, it cannot, in general, incentivize agents to report truthfully. Note that the impossibility result does hold if agents' beliefs satisfy the property of *stochastic relevance*, which states that agents with different observations have statistically different posterior beliefs (see Miller, Resnick, and Zeckhauser [15] for more details). This indicates the non-triviality of the obtained result, as under the same assumptions one can achieve truthful elicitation with a parametric mechanism, as described in the subsequent sections.

The impossibility result is especially important for elicitation of subjective beliefs, where agents beliefs can be skewed toward a particular observation due to the prior biases in the type of elicited information. We will see in the following sections how to establish truthful elicitation, first when agents' beliefs are fully known, and then when only prior beliefs are known, but agents' posterior beliefs are constrained by a belief updating rule. As shown by Frongillo and Witkowski [18], the latter can be done using different belief updating rule, but we focus in the book on one that is relatively easy to explain in terms of the maximum likelihood principle.

3.2 PEER CONSISTENCY FOR SUBJECTIVE INFORMATION

We now consider subjective information. Recall that subjective information arises when every agent observes a distinct sample drawn from the same distribution, as illustrated in Figure 3.1 on the right. For example, consider providing a review of a recent experience with Blue Star Airlines, widely considered to offer one of the best service in the sky. Unfortunately, the individual experience was not as good: the plane was delayed many hours, and the baggage was lost.

If we use the output agreement mechanism shown above, a rational agent would not report this poor service: after all, it is well known that almost everyone receives good service, so a peer agent is unlikely to match this poor experience.

The problem is that it is no longer appropriate to believe that the peer agent will report the same data, or, more formally, the agent's beliefs will no longer satisfy the self-dominating condition. So where could we anchor such a mechanism?

How does a poor experience affect an agent's belief? While she might not believe that the airline is all bad, we can still expect that the stellar image has received a dent, at least in this agent's belief, and that bad service will be considered more likely in the future than it was before the experience. We can use this change to construct incentives that make cooperative strategies optimal.

Types of peer consistency mechanisms for subjective tasks We use the characterization of beliefs and belief updates defined in Section 1.3. For objective information, we showed that output agreement induces a truthful equilibrium under the self-dominating condition (Definition 1.3) on agent beliefs, which is rarely violated in that setting. For subjective information, we require more restrictive conditions. The mechanisms that are known make one of one of the following two alternative assumptions.

- A homogeneous agent population with identical and known prior *and* posterior beliefs. An example is the peer prediction mechanism we show below.
- Common and known prior beliefs, but belief updates can be heterogeneous as long as they satisfy the *self-predicting* condition of Definition 1.4. An example we show below is the peer truth serum (PTS).

3.2.1 PEER PREDICTION METHOD

In the peer prediction method, introduced by Miller, Resnick, and Zeckhauser [15], we define for each observed value an *assumed* posterior distribution that reflects this shift (Figure 3.4).

Mechanism 3.2 Peer prediction.

---

1. Center gives a task to agents a_i; a_i reports data x_i.
2. Center randomly selects a *peer agent* a_j that has also been given the same task and reported data x_j.
3. Center selects an *assumed posterior* distribution $\hat{q}_{x_i}$ associated with report x_i.
4. Center pays agent a_i a reward:

$$pay(x_i, x_j) = SR(\hat{q}_{x_i}, x_j),$$

where *SR* is a proper scoring rule.

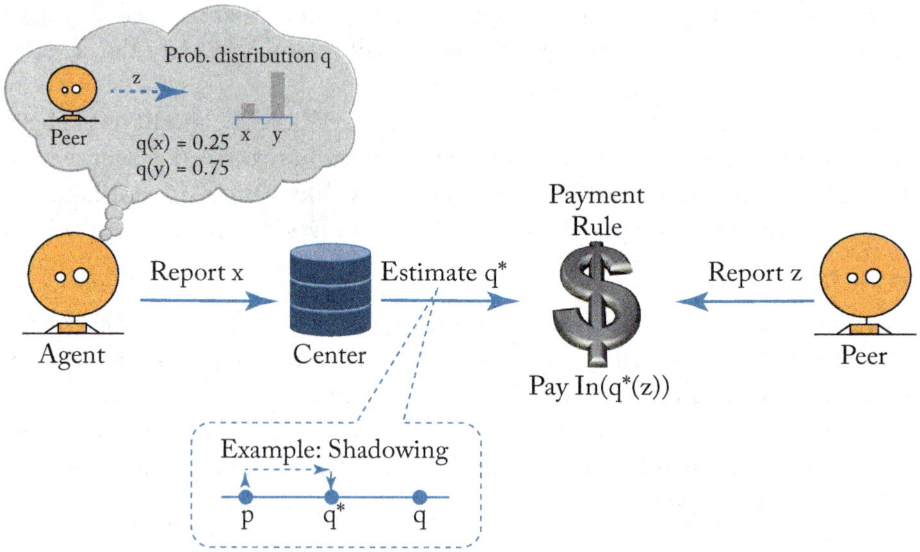

Figure 3.4: The peer prediction method.

The reward is calculated using a proper scoring rule as follows.

1. Each value for answer x_i is associated with an assumed posterior distribution $\hat{q}(x) = \hat{\Pr}(x|x_i)$.

2. $\hat{q}$ is skewed so that x_i is more likely than in the prior.

3. Use a proper scoring rule to score this posterior against a random peer report.

Thus, an agent implicitly reports its posterior probability distribution (provided, of course, that the distribution assumed by the platform is indeed its posterior). This allows rewarding even small shifts in the posterior beliefs. Note that the standard peer prediction procedure allows private information x_i to take real values, that is, $x_i \in \mathbf{R}$, as long as the assumed posterior $\hat{q}(x)$ is accurate, that is, it is equal to agents' posterior belief for the observation x_i.[1]

Let us see how this works in the example of the airline reviews. We assume that the agents report one of two values *good* and *bad*, and that they share a common prior distribution p:

$$p(good) = 0.\overline{8}(= 0.8888888...), p(bad) = 0.\overline{1}(= 0.111111\ldots)$$

and construct the assumed posterior distributions $\hat{q}$ for each answer using the mixture update (1.1) with $\delta = 0.1$:

[1] In the case of continuous x_i, the posterior belief $\hat{q}(x)$ is a probability density function.

Observation	$\hat{q}(good)$	$\hat{q}(bad)$
prior	0.8	0.1
good	$q_g(g) = 0.8 + \delta = 0.9$	$q_g(b) = 0.1$
bad	$q_b(g) = 0.8$	$q_b(b) = 0.1 + \delta = 0.2$

The payment function $pay(x, y)$ is a function of the report x and the peer report y. Assuming that we use a quadratic scoring rule:

$$2\hat{q}(x) - \sum_{x'} \hat{q}(x')^2$$

we obtain the following payment function:

		peer report	
		b	g
agent	b	$S_b(b) = 2 \cdot 0.2 - 0.68 = -0.28$	$S_b(g) = 2 \cdot 0.8 - 0.68 = 0.92$
report	g	$S_g(b) = 2 \cdot 0.1 - 0.82 = -0.62$	$S_g(g) = 2 \cdot 0.9 - 0.82 = 0.98$

We can check that an agent who observed bad service now has reason to report this fact truthfully. The assumed posterior for reporting bad service is $\hat{q}_b(g) = 0.8$:

$$E[pay(\text{``}bad\text{''})] = 0.2 \cdot \underbrace{pay(b, b)}_{=-0.28} + 0.8 \cdot \underbrace{pay(b, g)}_{=0.92} = 0.68$$

$$E[pay(\text{``}good\text{''})] = 0.2 \cdot \underbrace{pay(g, b)}_{=-0.62} + 0.8 \cdot \underbrace{pay(g, g)}_{=0.98} = 0.66.$$

Thus, we have made reporting the correct value profitable, even when it is not the most likely answer!

In general, we can show for the mechanism:

Theorem 3.4 *The Mechanism 3.2 has a strict Bayes-Nash equilibrium where all agents use cooperative strategy, provided that all agents have the common beliefs and belief updates assumed in the mechanism.*

The proof follows in a straightforward way from the fact that proper scoring rules yield the highest expected reward when the correct distribution is used, and the only way for agents to influence this distribution is to report their believed values as accurately as possible.

What is the revenue of the peer prediction mechanism for agents that report randomly according to their prior distribution? This depends very much on how the assumed posteriors are constructed.

If we assume that both mechanism and agent compute the posterior distributions using the Bayesian update as in Equation (1.1), we obtain for the quadratic scoring rule the following expected payment for reporting x:

$$\begin{aligned} E[pay(x)] &= 2(p(x)(1-\delta)\delta) - \sum_y [p(y)(1-\delta]^2 + [p(x)(1-\delta)]^2 - [p(x)(1-\delta) + \delta]^2 \\ &= (1-\delta)^2 \left(2p(x) - \sum_y p(y)^2 \right) + \delta - \delta^2 \end{aligned}$$

and thus the expected payment overall, given the prior distribution P:

$$\begin{aligned} E[pay] = \sum_x p(x) E[pay(x)] &= (1-\delta)^2 \left[2 \sum_x p^2(x) - \sum_x p(x)^2 \right] + \delta - \delta^2 \\ &= (1-\delta)^2 \sum_x p^2(x) + \delta - \delta^2 \end{aligned}$$

which for small δ is proportional to $\lambda(P) = \sum_x p(x)^2$. In the example above, $n = 2$ and $\lambda(P) = 0.80$, and with $\delta = 0.1$, we would obtain $E[pay] = 0.738$. This is the value that has to be subtracted from the payments to eliminate rewards for reporting according to the prior. Note that for an agent who receives bad service, the rewards of the final scheme are always significantly negative, no matter what value is reported. The high volatility of payments can be a big obstacle to practical use of the scheme.

For the logarithmic scoring rule, a similar derivation shows convergence to $-H(P)$ for small δ.

3.2.2 IMPROVING PEER PREDICTION THROUGH AUTOMATED MECHANISM DESIGN

The original construction of peer prediction through the use of a proper scoring rule was a significant breakthrough, but the solution has two problems. The first is that general proper scoring rules generate inefficient and volatile payments. In the example above, the gain from truthfulness is only 0.02 on a payment of 0.68, so to compensate a measurement cost of \$1 would require payments of at least \$34, which is a huge premium to be paid for truthful elicitation.

The second problem is that the mechanism has other, more profitable but uninformative equilibria: always reporting "good" gives the much higher expected payoff of 0.98. In fact, it is possible to show that any mechanism based on just 2 reports will always have uninformative equilibria with a higher payoff than the truthful one [16].

Both problems can be solved using a technique called *automated mechanism design*, as we will show next. The inefficiency can be addressed by automatically designing the entries of the payment function $pay(x, y)$ by solving a linear program. The program can choose payments that minimize the expected payments. The problem of uninformative equilibria can be avoided by scoring reports against the distribution of multiple peer reports, using the same technique of automated design through a linear program.

Efficient payments To simplify the mechanism, we assume that there is no payment for reports that do not agree. To ensure that cooperative, truthful strategies form an equilibrium, we need to find payments $pay(g, g)$ and $pay(b, b)$ such that:

$$\begin{aligned} q(g|g)pay(g, g) &> q(b|g)pay(b, b) + \epsilon_g \\ q(b|b)pay(b, b) &> q(g|b)pay(g, g) + \epsilon_b, \end{aligned}$$

where ϵ_g and ϵ_b are the minimum differences in reward we want to achieve for reporting truthfully "good" and "bad," and we assume that we only pay for reports that agree with the peer (i.e., $pay(g, b) = pay(b, g) = 0$).

In this example, assuming $\epsilon_g = \epsilon_b = 0.1$:

$$\begin{aligned} 0.9pay(g, g) &> 0.1pay(b, b) + 0.1 \\ 0.2pay(b, b) &> 0.8pay(g, g) + 0.1. \end{aligned}$$

These constraints define a space of feasible payments shown in Figure 3.5.

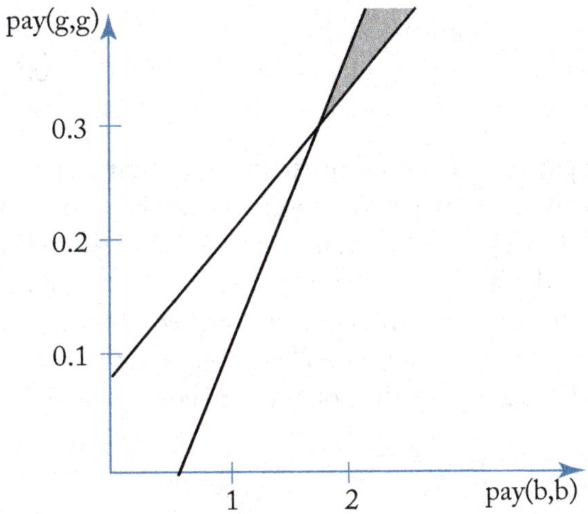

Figure 3.5: Linear program for determining the optimal payments that support a truthful equilibrium.

The optimization function is to minimize the expected expenditure, given that the data can be assumed distributed according to the prior distribution:

$$\text{Minimize } p(g)q(g|g)pay(g, g) + p(b)q(b|b)pay(b, b) = 0.8pay(g, g) + 0.0\bar{2}pay(b, b).$$

Thus, we obtain as a solution the payments in the lower corner of the feasible region in Figure 3.5:

$$pay(g, g) = 0.3, pay(b, b) = 1.7.$$

The expected payment is $0.2\overline{7}$, where reporting truthfully is always better than not by a margin of 0.1; thus, to compensate a measurement cost of \$1, we need a payment of at least \$2.78, which is not such a big markup for guaranteeing the quality.

Uninformative equilibria The mechanism we just constructed has three pure-strategy equilibria:

1. truthful: expected payment = $0.2\overline{7}$,
2. always reporting "good": expected payment = 0.3, and
3. always reporting "bad": expected payment = 1.7.

Clearly, agents would prefer to always report "bad," leaving the center with no information about the true value. Consider now scoring against not one, but several reference reports. For example, we could use 3 reference reports and count the number of "good's" among them. Thus, we obtain the following 8 situations with their associated probabilities:

$\Pr(\lvert peer = g \rvert \lvert o)$	0	1	2	3
$o = b$	0.008	0.096	0.384	0.512
$o = g$	0.001	0.027	0.243	0.729

and we have to define a payment for each of them, such that the expected value of reporting *good* exceeds that of reporting *bad* when the observation o is *good*, and vice versa for a *bad* observation. Together with the objective function of minimizing the expected payment, this again defines a linear program to design the mechanism. However, we can add additional constraints. In this case, to eliminate the uninformative pure-strategy equilibria, we can simply force the corresponding reward to (for all "bad" or all "good" reports) to zero, and to a small positive reward ϵ for deviating from this situation. We thus obtain a payment function such as:

$pay(r, \lvert peer = g \rvert)$	0	1	2	3
$r = b$	0	10	0	ϵ
$r = g$	ϵ	0	2	0

where we note that truthtelling is a strict equilibrium:

$$\begin{aligned} o = bad: E[pay(\text{``}bad\text{''})] = 0.96 &\quad > \quad E[pay(\text{``}good\text{''})] = 0.768 \\ o = good: E[pay(\text{``}bad\text{''})] = 0.27 &\quad < \quad E[pay(\text{``}good\text{''})] = 0.468 \end{aligned}$$

but all "*good*" or all "*bad*" is not a strict or weak equilibrium.

Kong, Ligett, and Schoenebeck [17] show how to modify scoring rules so that the truthful equilibrium is guaranteed to have the highest payoff not only among pure strategies (as assumed in the construction above), but also among mixed strategies. The construction is more complex to follow so we do not describe it in more detail here.

3.2.3 GEOMETRIC CHARACTERIZATION OF PEER PREDICTION MECHANISMS

Frongillo and Witkowski [18] showed that there is a unique mapping between any truthful payment rule for a peer prediction mechanism and a corresponding proper scoring rule. Thus, even mechanisms designed using automated mechanism design can be understood as derived from proper scoring rules in the same way as the original peer prediction mechanism.

The construction is based on considering the expected payments for reporting a value x:

$$E[pay(x)|o] = \sum_y q(y|o)\tau(x, y)$$

as a function of the posterior distribution $q(x|o)$. For each possible posterior distribution $q(x|o)$, there is a value x that gives the highest expected reward and that a rational agent would report.

We can now consider that the space of all possible posterior distributions is broken into cells where a particular value is the optimal report. It turns out that because of the linearity of the expectation, this division is that of a *power diagram*. The cells are defined by the areas around the n points $\underline{v}^x$ defined by the vectors of payments $(\tau(x, x_1), \tau(x, x_2), \ldots, \tau(x, x_n))$ that are obtained when reporting the value x. Note that the expected payment for reporting x:

$$E[pay(x)] = \underline{q} \cdot \underline{v}^x$$

is maximized when $\underline{q} = \frac{1}{\sqrt{w(x)}}\underline{v}^x$, where $w(x) = ||\underline{v}^x||$. If there is a non-empty set of posterior distributions where reporting x yields the optimal payment, then this distribution is one of them. Furthermore, due to the linearity of the payment function, the same holds for a neighborhood of similar posterior distributions that we call the *cell* associated with v^x. Frongillo and Witkowski [18] show that this cell is characterized by the distributions u such that v^x has the lowest *power distance*:

$$||u - v^x||^2 - w(x).$$

Figure 3.6 illustrates the construction for the example of good and bad service shown earlier, and the payment rule derived using automated mechanism design. We have two values: g(ood) and b(ad), and so the diagram has two dimensions $(q(g), q(b))$ that characterize the agent's possible posterior beliefs. However, since the points form probability distributions, the space actually has only one dimension ϕ, characterizing the probability of good service, and distributions are characterized as $q = (\phi, 1 - \phi)$. This space is shown as a thick line. The sites are $v^b = (0, 1.7)$ and $v^g = (0.3, 0)$, and the power distance to the site "g", with weight $w(g) = 0.3^2$ is:

$$(\phi - 0.3)^2 + (1 - \phi)^2 - 0.3^2 = 2\phi^2 - 2.6\phi + 1$$

while the power distance to the site "b" with weight $w(b) = 1.7^2$ is:

$$\phi^2 + (1 - \phi - 1.7)^2 - 1.7^2 = 2\phi^2 + 1.4\phi - 2.4.$$

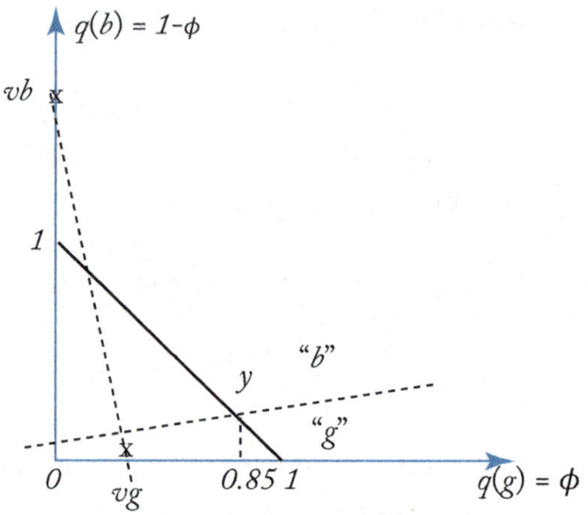

Figure 3.6: Power diagram construction for the example.

For any of the possible agent posterior beliefs q on the admissible line, the value that will give the higher expected payoff is the one of the site with the smallest power distance. Both distances are equal for all points on the line that is perpendicular to the line that connects v^g and v^b and passes through the point y at $(0.85, 0.15)$, shown as a dashed line in Figure 3.6. This line forms two cells, and their projection onto the admissible line of distributions gives intervals of $\phi \in [0, 0.85)$ where "b" gives a better payoff, and $\phi \in (0.85, .1]$ where "g" gives a better payoff. The power diagram construction can be carried out in higher-dimensional spaces and thus deal with scenarios with more than two values.

The pivotal point in the power diagram construction is the point y where all cell boundaries meet. For a posterior distribution congruent with y, the mechanism gives an equal reward to all values. Ideally, this should be the prior distribution, so that an agent that does not make the effort to observe the phenomenon does not have a strategy that pays a reward.

Frongillo and Witkowski [18] show that any peer prediction mechanism can be characterized by such a power diagram, and suggest that this relationship can be exploited for designing such mechanisms. Here is a slightly modified version of their procedure, that takes an estimated prior probability distribution y as an input and ensures that the mechanism is ambivalent for this distribution.

1. Starting with an assumption of how agents might form their posterior beliefs based on an observation, one infers the shape of the cells in the power diagram where it results from a particular observation—in the example, the assumption that agents increase the posterior probability of the value they observed leads to the split along the prior probability at 0.85. Furthermore, the prior probability distribution defines a central point $\underline{y} = (0.15, 0.85)$.

2. Pick an arbitrary point in one of the cells, say for value x_1, as the site v_1 for this cell. In the example, we might pick b and $v_b = (1.7, 0)$. The site fixes a tentative payment rule for agents that report x_1: the coordinates fix the payments for the different possible peer reports.

3. Pick a site for a neighboring cell, say for x_2, such that the difference is the vector $\underline{u}$ perpendicular to the cell boundary between them, characterized by the condition $\underline{u} \cdot \underline{y} = 0$. In the example, this would be for example the vector $(-0.85, 0.15)$, and so we could choose $v_g = (0, 0.3)$.

4. Iterate: pick a neighboring cell for value x_k and choose its site to lie on the intersection of lines that pass through the sites already chosen and that are perpendicular to the boundaries between the respective cells (see Frongillo and Witkowski [18] for more details). In the example, there are no further values so no further sites have to be found.

5. Set the weights: all sites must have equal power distance to the central point y, and so the weights of the sites might have to be adjusted. In the example, this is not required, as the points are already chosen to satsify this condition.

The mechanism is given by the payment rules corresponding to the coordinates of the sites. Usually, they should be rescaled to optimize budget or other constraints, and ensure that reports according to the prior give an expected reward of zero.

3.3 COMMON PRIOR MECHANISMS

3.3.1 SHADOWING MECHANISMS

Clearly, a big weakness of the peer prediction method is that it requires the mechanism designer to know the exact posterior distributions formed by an agent for each different observation. Even worse, since one and the same mechanism has to apply to all agents, these distributions have to be the same for every agent that participates in the mechanism! Such a strict condition is unlikely to hold, and so we should look for a way to weaken it.

It is important to observe that in most scenarios, it would actually be quite reasonable to assume that agents have a common *prior* belief before they observe the phenomenon. For example, our prior belief about the quality of a product could reasonably be formed by the reviews published so far, our prior belief about the temperature would be the historical average, and we may consider all answers to a crowdworking task to be equally likely. Furthermore, this prior may actually be known to the center with reasonable precision.

However, agents differ a lot in the way they form their posterior beliefs: they have different confidence in their own observations, their observations may differ, and so on. Witkowski and Parkes [19] observed that when constructing the assumed posterior distributions, the exact value of δ does not matter (for two values and the quadratic scoring rule): as long as an agent increases $q(bad)$ over the prior when observing "bad," the expected reward $E[pay(\text{``}bad\text{''})]$ is strictly greater than $E[pay(\text{"good"})]$ for *any* $\delta > 0$ that satisfies some well-formedness constraints of the distribution.

They were thus able to show that constructing the assumed posteriors by such *shadowing* allowed the peer prediction principle to work even when agents differ in their posterior beliefs. However, they used this observation in the context of a Bayesian Truth Serum construction (see Chapter 4). The first version of the construction worked only for binary signals.

The construction using power diagrams we showed above clearly illustrates that this observation will hold more generally: the same mechanism will be truthful as long as agents' posterior beliefs fall into the right cells. Furthermore, when their posterior beliefs are derived by incremental updates such as shadowing from the prior beliefs, the central point of the cells will correspond to a common prior belief.

3.3.2 PEER TRUTH SERUM

We can go further on the shadowing idea and bypass the explicit construction of a posterior distribution entirely, only relying on the presence of a common prior. This will give us a simple mechanism that works for any number of values, which we call the *peer truth serum*.

In the construction, we apply the shadowing idea in a slightly different way; instead of linearly increasing/decreasing the probability of the observed value by δ, we consider the frequentist belief update introduced in Equation (1.1) from the common prior p, for an observed value x_i:

$$\begin{aligned}\hat{q}(x_i) &= p(x_i) + (1 - p(x_i)) \cdot \delta = \delta + p(x_i) \cdot (1 - \delta) \\ \hat{q}(x_j) &= p(x_j) \cdot (1 - \delta) \text{ for } x_j \neq x_i\end{aligned}$$

so that the derivative of $\hat{q}$ with respect to the parameter δ is as follows:

$$\frac{d\hat{q}(x)}{d\delta} = \begin{cases} 1 - p(x) & x = x_i \\ -p(x) & x = x_j \neq x_i \end{cases} = \mathbf{1}_{x=x_i} - p(x).$$

Our construction will use the logarithmic scoring rule:

$$LSR(\underline{q}, g) = C + \ln q(g)$$

where we assume $C = 0$, to score a peer report x_p against the assumed posterior beliefs:

$$LSR(\underline{\hat{q}}, x_p) = \ln \hat{q}(x_p)$$

Alternatively, we can also interpret the assumed posterior beliefs as the model that the center obtains after updating according to x_i. The frequentist update would correspond for example to the center maintaining a histogram of reports.

Rather than applying the scoring rule explicitly, we use an approximation by its Taylor expansion with respect to δ for the *improvement* of the prediction of the peer report by the updated model $\underline{\hat{q}}$ in comparison with the prior $\underline{p}$.

The derivative of the log scoring rule is:

$$\frac{\partial LSR(\underline{p}, x_p)}{\partial p(x)} = \begin{cases} 1/p(x) & x = x_p \\ 0 & x \neq x_p \end{cases}$$

and so the Taylor expansion of the logarithmic scoring rule applied to the assumed distribution is as follows. Since we would like random reporting according to the prior distribution $\underline{p}$ to be equal to 0, we make this the starting point of the expansion. We can then write the payment for a peer report x_p as:

$$\begin{aligned}
LSR(\underline{\hat{q}}, x_p) - LSR(\underline{p}, x_p) &\approx \delta \cdot \frac{d\, LSR(\underline{\hat{q}}, x_p)}{d\delta} \\
&= \delta \sum_z \frac{\partial LSR(\underline{\hat{q}}, x_p)}{\partial \hat{q}(z)} \frac{d\hat{q}(z)}{d\delta} \\
&= \delta \sum_z \left(\frac{\mathbf{1}_{z=x_p}}{p(z)} \right) \left(\mathbf{1}_{z=x_i} - p(z) \right) \\
&= \delta \left(\sum_z \frac{\mathbf{1}_{z=x_i} \mathbf{1}_{z=x_p}}{p(z)} - \sum_z \mathbf{1}_{z=x_p} \frac{p(z)}{p(z)} \right) \\
&= \delta \left(\frac{\mathbf{1}_{x=x_p}}{p(x)} - 1 \right)
\end{aligned}$$

We thus obtain expressions that are linear in δ, and in fact δ just becomes a scaling factor that does not influence the qualitative character of encouraging cooperative behavior in the agents!

Mechanism 3.3 The peer truth serum (PTS).

1. Center informs all agents of prior distribution R used by the mechanism.
2. Agents a_i carries out a task and observes value o; a_i reports data x_i.
3. Center randomly selects a *peer agent* a_j that has also been given the same task and reported data x_j.
4. Center pays agent a_i a reward:

$$pay(x_i, x_p) = \frac{\mathbf{1}_{x_i=x_p}}{r(x_i)} - 1.$$

We thus obtain the *peer truth serum*, shown in Mechanism 3.3. It has as main parameter a distribution R that is known to the center, and rewards agreement of report x_i with a peer report x_p with

$$pay(x_i, x_p) = \frac{\mathbf{1}_{x_i = x_p}}{r(x_i)} - 1. \tag{3.1}$$

We can see that this payment scheme *by construction* aligns incentives between the reporting agents and the model learning process of the center. Under what condition does this scheme induce truthful, cooperative strategies? First assume that the agent prior P is equal to R. We write the incentive compatibility condition:

$$\begin{aligned} & E_{Q(x|x_i)}[pay(x_i, x)] = q(x_i|x_i) \cdot pay(x_i, x_i) = q(x_i|x_i)/r(x_i) \\ > \quad & E_{Q(x|x_i)}[pay(x_j, x)] = q(x_j|x_i) \cdot pay(x_j, x_j) = q(x_j|x_i)/r(x_j) \end{aligned}$$

and note that when $R = P$, this translates to the *self-predicting* condition introduced in (1.3):

$$\frac{q(x_i|x_i)}{p(x_i)} > \frac{q(x_j|x_i)}{p(x_j)}, i \neq j$$

or, equivalently, x_i must be the maximum-likelihood estimate:

$$q(x_i|x_i) > q(x_i|x_j), i \neq j.$$

Defining admissible beliefs as a common prior beliefs and belief updates that satisfy the self-predicting condition, we can show the following theorem.

Theorem 3.5 *For self-predicting belief updates, the Peer Truth Serum as shown in Mechanism 3.3 has a strict ex-post subjective Bayes-Nash equilibrium where all agents report truthfully.*

The proof is straightforward given that we have shown the self-predicting condition as the sufficient condition for truthful reporting to be the best response strategy in the PTS mechanism. Note that the self-predicting condition is satisfied for example when agents use frequentist belief updates according to Equation (1.1), but not when using a mixture update (which can be handled by an alternative version based on the quadratic scoring rule, shown below).

Clearly, agents' beliefs can satisfy this condition to different degrees, reflecting the confidence. Let us characterize this confidence γ of agent a as:

$$\gamma_a(x_i) = \frac{q(x_i|x_i)}{p(x_i)} - 1.$$

Given the payment function (3.1), and assuming that $P = R$, the expected revenue of an agent with a cooperative strategy reporting x_i is just equal to its expected confidence $\gamma_a(x_i)$

$$E[pay] = E[\gamma_a] = \sum_{x_i} p(x_i)\gamma_a(x_i).$$

Thus, the payment increases linearly with confidence, and the PTS scheme motivates agents to invest effort to increase their confidence in the data. Another observation relates to the granularity of measurement. As the expected confidence is capped by the fact that the posterior probability cannot exceed 1, observing a phenomenon with few values or a skewed prior distribution has a lower potential for rewards than one with many values and an even prior distribution. This is important for motivating agents to observe more complex phenomena (see Chapter 8). The analysis here has made the simplifying assumption that $R = P$; see Faltings et al. [22] for a complete analysis without this assumption.

We further define the agent's *self-predictor* as the smallest Δ_a such that:

$$\Delta_a \left(\frac{q(x_i|x_i)}{p(x_i)} - 1 \right) > \frac{q(x_j|x_i)}{p(x_j)} - 1, \forall x_i, x_j, x_i \neq x_j. \tag{3.2}$$

Δ_a is a number in $[0, \ldots, 1]$ that characterizes how much values are correlated to one another. If they are categorical, i.e., no value is positively correlated to another, it is equal to 0. On the other hand, if there is a pair of values that are perfectly correlated (so that they are indistinguishable), it is equal to 1. $1 - \Delta_a$ can be understood to characterize the efficiency of the mechanism. It is equal to the fraction of the expected payment that is given for truthful reporting. The self-predictor will also be important in the properties of a variant of PTS, the peer truth serum for crowdsourcing, that we present in Chapter 5.

We also note that the PTS mechanism that we derived here from the logarithmic scoring rule can also be derived for other scoring rules. Let us consider the analogous derivation to the one shown above for the quadratic scoring rule. Consider thus a payment function:

$$pay(\underline{p}, x_p) = 2p(x_p) - \sum_y p(y)^2$$

with the derivative:

$$\frac{\partial pay(\underline{p}, x_p)}{\partial p(x)} = -2p(x) + \begin{cases} 2 & x = x_p \\ 0 & x \neq x_p \end{cases} = 2\left(1_{x=x_p} - p(x)\right).$$

The Taylor expansion from the prior distribution $\underline{p}$ is:

$$
\begin{aligned}
QSR(\underline{\hat{q}}, x_p) - QSR(\underline{p}, x_p) &\simeq \delta \cdot \frac{d\,pay(\underline{\hat{q}}, x_i)}{d\delta} \\
&= \delta \sum_z \frac{\partial pay(\underline{p}, x_i)}{\partial p(z)} \frac{d\hat{q}(z)}{d\delta} \\
&= 2\delta \sum_z \left(\mathbf{1}_{z=x_p} - p(z)\right)\left(\mathbf{1}_{z=x_i} - p(x_i)\right) \\
&= 2\delta \left(\sum_z \mathbf{1}_{z=x_i}\mathbf{1}_{z=x_p} - p(x_i) \underbrace{\sum_z \mathbf{1}_{z=x_p}}_{=1} - \sum_z \mathbf{1}_{z=x_i} p(z) + p(x_i) \underbrace{\sum_z p(z)}_{=1} \right) \\
&= 2\delta \left(\mathbf{1}_{x_i=x_p} - p(x_i)\right).
\end{aligned}
$$

This payment rule is incentive-compatible under a slightly different version of the self-predicting condition:

$$q(x_i|x_i) - p(x_i) > q(x_j|x_i) - p(x_j)$$

which is incomparable to the maximum-likelihood condition that results for the logarithmic scoring rule but may be more suitable in some scenarios. Note that it is satisfied for example when agents update their beliefs according to the Bayesian update given in Equation (1.1) for a single value, but not always for the Bayesian update in Equation (1.4).

Likewise, similar derivations can be made for any proper scoring rule, although they might not always result in simple expressions for the rewards.

Helpful reporting What happens if the R assumed by the center is different from the agents' prior distribution P?

When the center has little information about the data to be elicited, it may not have a good idea of the prior distribution of the agents that collect the information, and so the R used in the payment rule may differ from the P that characterizes agents' common prior distribution, and thus does not encourage truthful reporting.

However, in such a case, it will be reasonable to assume that the agents' prior P over/under-estimates R whenever P^* under/over-estimates R, a property that we call *informed*.

Definition 3.6 Probability distribution P is *informed* with respect to distribution R and true distribution P^* if and only if for all values x, $(r(x) - p^*(x))(r(x) - p(x)) \geq 0$.

In this case, agents partition values into two sets:

- under-reported values, with $r(x) < p(x)$: due to informedness, for these values we also have $r(x) < p^*(x)$, and

- over-reported values, with $r(x) \geq p(x)$, and where informedness implies $r(x) \geq p^*(x)$

and this partitioning is the same for all agents due to the common prior belief.

If an agent adopts a non-truthful strategy to report a value x instead of y, we may find one of the following two situations:

- the strategy may be profitable if x is under-reported or y is over-reported, since the center under-estimates the correct reward for x and over-estimates it for y; but
- it is never profitable if x is over-reported and y is under-reported, since the agent would expect a smaller reward from mis-reporting.

Thus, an agent whose prior beliefs are infored with respect to the distribution R will adopt a *helpful* reporting strategy, defined as follows [20].

Definition 3.7 A reporting strategy is *helpful* if it never reports an over-reported value x for an under-reported value y.

We will show in Chapter 8 that when the center updates the distribution R using the Bayesian, frequentist update model (Equation (1.1)), such helpful reporting guarantees a property called *asymptotic accuracy*.

Properties of the peer truth serum We can show several interesting properties of the Peer Truth Serum. First of all, it is unique: any payment function that incentivizes truthful reporting with only the self-predicting condition must have the form $f = 1/p(x_i) + g(-x_i)$, where $g(-x_i)$ is a function independent of the report x_i. Because of this uniqueness, we can also show that it is *maximal*, and that weakening any of the assumptions makes truthful incentive mechanisms impossible.

In particular, the following theorem holds.

Theorem 3.8 *There does not exist an* asymptotically accurate mechanism *that has truthful reporting as a* strict ex-post subjective Bayes-Nash equilibrium *for a general structure of agents' beliefs.*

The take away message of this impossibility is that one cannot trivially relax the self-predicting condition. Notice that Theorem 3.8 does directly follow from Theorem 3.3, as we are now allowing mechanisms to depend on parameter.

Finally, the peer truth serum incentivizes *optimal* information gathering: when the loss function is the logarithmic scoring rule, the agents are incentivized to report in a way that most reduces this loss function for the estimate constructed by the center. We will discuss this in more detail in Chapter 8.

Other equilibria As in the output agreement mechanism, cooperative, truthful strategies are not the only equilibrium. In fact, it is easy to see that in the peer truth serum, the equilibrium with the highest possible payoff is for all agents to report the x with the smallest $r(x)$. This will lead to an uniformative, uniform distribution that will eventually reduce the payoff of any strategy to 0, but we cannot exploit this feature to eliminate it. However, if employed by a significant number of agents, it can be easily detected, as one would observe many agents simultaneously reporting the same unlikely value, and this value would be different in different time intervals. Thus, we can adopt a similar solution as in the ESP game: penalize all agents when an uninformative equilibrium is detected.

A more elegant solution to eliminate this possibility is to not publish distribution R, but derive it from multiple answers. We will show such a mechanism in Chapter 5.

3.4 APPLICATIONS

We now present several applications of peer prediction and the peer truth serum that have been presented in the literature. We stress that these are mostly simulated; the only existing applications of crowdsourcing to data acquisition are in human computation platforms such as Amazon Mechanical Turk and usually lack incentive mechanisms.

3.4.1 PEER PREDICTION FOR SELF-MONITORING

Providers of services, such as internet access, mobile phone service, or cloud computing services, operate under Service Level Agreements (SLA). These agreements stipulate penalties for insufficient quality of service. However, monitoring and proving insufficient quality is costly, and it would actually be best if it could be done by users themselves. However, since they can claim refunds or other penalties if the conditions of the SLA are not met, they have a natural incentive to incorrectly report poor service.

Here, we can use the peer consistency idea to make truthful reporting of the actual quality of service an optimal strategy [21], and thus allow self-monitoring. Clearly, such a mechanism is prone to uninformative equilibria where agents report poor service, and will only work under the assumption that the size of coalitions that simulteneously report poor service can be limited. This can be the case for example when the agents have to report more details of the service they receive at many time points, and there are thus many uninformative equilibria that agents will have difficulty to coordinate on. The mechanism is intended for cases where outages are sporadic and cases where a large fraction of agents receives poor service do not need to be caught as they will be detected by other means as well.

For example, consider a web service provider that delivers a data service (for example, a weather forecast) to a homogeneous population of users. Assume the service has two quality parameters:

- $Q_1(0/1)$: whether the response was received within the stipulated response time, and

- $Q_2(0/1)$: whether the provided information is correct.

Assume, furthermore, that the cost the service provider incurs for providing the service is equal to \$1.0 per report, that the refund each *user* gets in case of a poor report is \$0.01, and that the synchronization cost of user misreporting is \$10.00 for any number of reports.

We can design incentives according to peer prediction that balance the incentive to misreport (\$0.01). Assuming that the prior probability distributions for the 2 quality indicators Q_1 and Q_2 are 0.9 for "1" (good) and 0.1 for "0" (bad), and that the posterior probability changes by 20%:

$$\begin{aligned} \hat{q}(1|1) &= 0.92, \hat{q}(1|0) = 0.88 \\ \hat{q}(0|0) &= 0.12, \hat{q}(0|1) = 0.08 \end{aligned}$$

then the expected cost of the incentives is as shown in Figure 3.7. The cost can, however, be much reduced if there are some agents whose reports are absolutely trustworthy and that can be used as peers to eliminate uninformative equilibria.

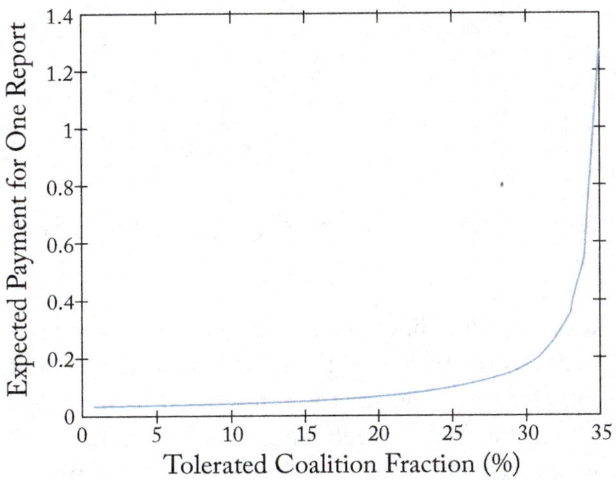

Figure 3.7: Monitoring cost using peer prediction as a function of the maximum fraction of colluding agents.

3.4.2 PEER TRUTH SERUM APPLIED TO COMMUNITY SENSING

Faltings et al. [22] evaluated the performance of the peer truth serum on the community sensing application we presented in the introduction. The study shows how agents can be incentivized by peer consistency without the need for any ground truth.

A major challenge here is that the mechanism requires peer reports, but there are never multiple sensors that measure at the exact same location. Thus, we need to either use a sensor that is close as a peer—a simple solution but we cannot expect exactly the data to completely agree—or we use a *pollution model* to predict what data a peer *should* observe based on all peer reports that enter the model in the same time period. This allows applying the peer truth serum to reward sensor owners for their cost of operating the sensor in the most accurate way, as shown in Faltings et al. [22].

The basis is a simulation model of NO_2 constructed by environmental scientists for the city of Strassbourg (France) (Figure 3.8). Measurements are discretized to a scale of three values. The model is based on measurements taken by high-quality sensors over a period of four weeks, and interpolation to other simulated sensors using a numerical pollution propagation model.

Figure 3.8: Pollution model of the city of Strassbourg. The red crosses show the 116 sensor locations. Courtesy of Jason Li.

Simulations based on this data have been used to evaluate the peer truth serum (and also other techniques shown later in the book). In the simulations, agents observe the values as given in the dataset with varying degrees of noise. The center used Gaussian process regression based on the other data obtained so far to derive reference measurements for the points where agents reported data. This setup allowed for simulating different agent strategies and quantifying their impact on the model learned by the center.

The first question is whether the game-theoretic properties hold up in the presence of measurement noise, which can be quite substantial with low-cost sensors. Figure 3.9 shows the performance of three different strategies as a function of the noise level. The blue curve shows the average reward when always reporting truthfully the (noisy) simulated measurement, compared to the strategy of reporting randomly according to the prior distribution, and always reporting the lowest possible value.

Another interesting question is whether the incentive scheme encourages agents to place their sensors at locations where their measurement contributes a lot of insight. Given an incentive scheme where payment is for agreeing with others, an agent can minimize her risk of disagreement with a peer by choosing locations where measurements are very certain. This can be

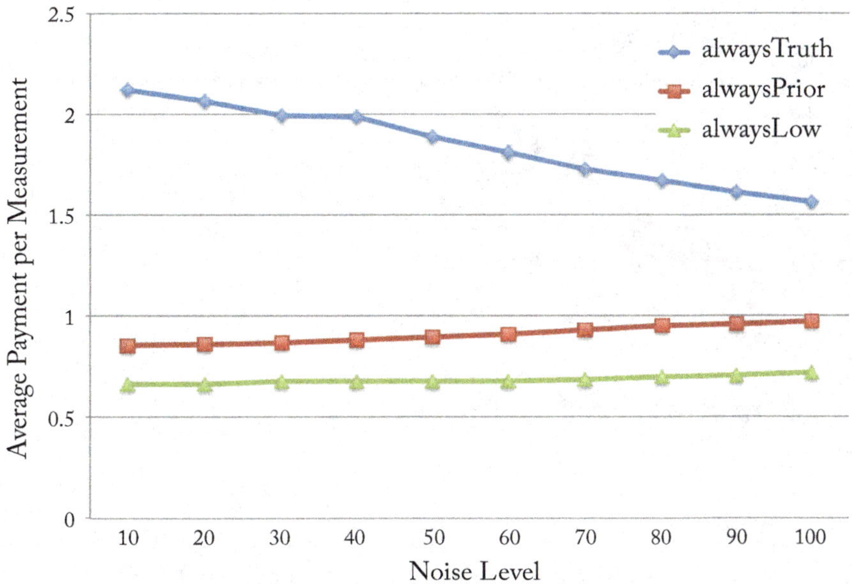

Figure 3.9: Rewards for different strategies as a function of measurement noise (standard deviation as percentage of average value). Courtesy of Jason Li.

observed for example when rewarding measurements according to a proper scoring rule (Chapter 2). As Figure 3.10 shows, the peer truth serum tends to encourage measurements at locations with higher uncertainty, even though this was not a design objective.

The big issue with the peer truth serum is of course whether the scheme is vulnerable to uninformative equilibria where agents collude to report either always the same value, or the value with the smallest value of $r(x)$ and thus the highest payment. The collusive strategies are complicated somewhat by the fact that information is aggregated into the Gaussian model.

Figure 3.11 shows the average rewards for several different strategies in a situation where a significant coalition of agents collude to report the value that is currently the least likely; this is the most profitable uninformative equilibrium. We can see that strategies of always reporting the lowest value, or reporting according to the prior distribution are not interesting. The truthful strategy remains best up to a coalition size of over 60% of the agents, which means that the scheme is actually quite robust. Even more so, when the coalition of colluders is below 40%, it obtains almost no reward, so it's difficult to motivate agents to join.

3.4.3 PEER TRUTH SERUM IN SWISSNOISE

In order to understand the design and behavior of public prediction markets, Garcin and Faltings [55] have implemented a public platform called *Swissnoise*. It was operated from EPFL

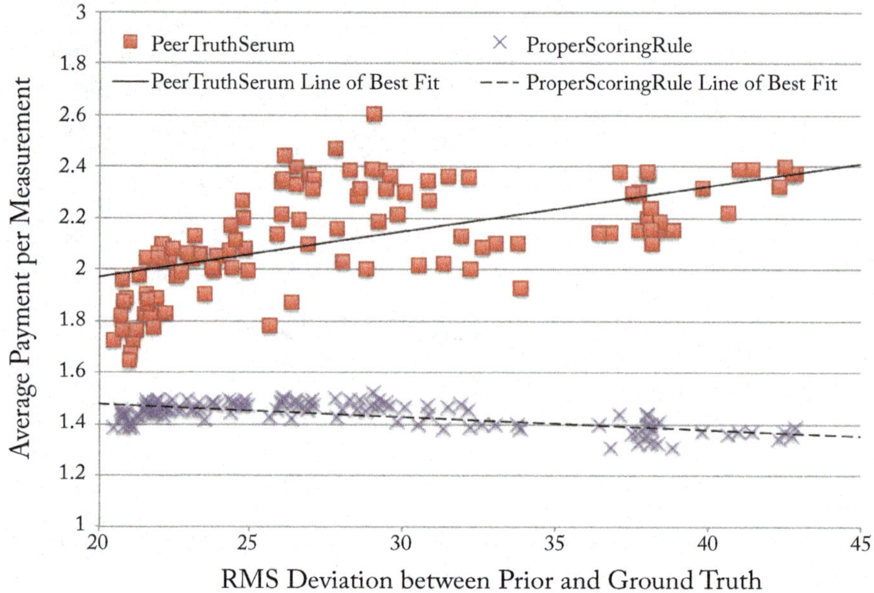

Figure 3.10: Expected reward as a function of the uncertainty of the measurement, as expressed by the RMS deviation between the prior distribution and the actual value as given by the pollution model. Courtesy of Jason Li.

during from spring of 2013 to the summer of 2015 with up to 300 participants, and allowed predictions on questions of current public interest, usually suggested by participants themselves. It was operated with artificial money as a prediction market with a logarithmic scoring rule market maker. The artificial money accumulated by each participant was shown in a leaderboard, and each week the participant with the biggest profit was awarded a small gift certificate.

During its operation, Swissnoise handled more than 230 questions and 19,700 trading operations were carried out. Questions were of different kinds, including:

- sports events, such as the FIFA soccer world cup in 2014;
- political events, such as the outcome of the referendum on Scottish independence in 2014, and the outcomes of numerous Swiss referendums;
- technological milestones, such as when the first Chinese lunar lander would reach the moon; and
- local events concerning the EPFL campus, such as whether the local grocery store would close.

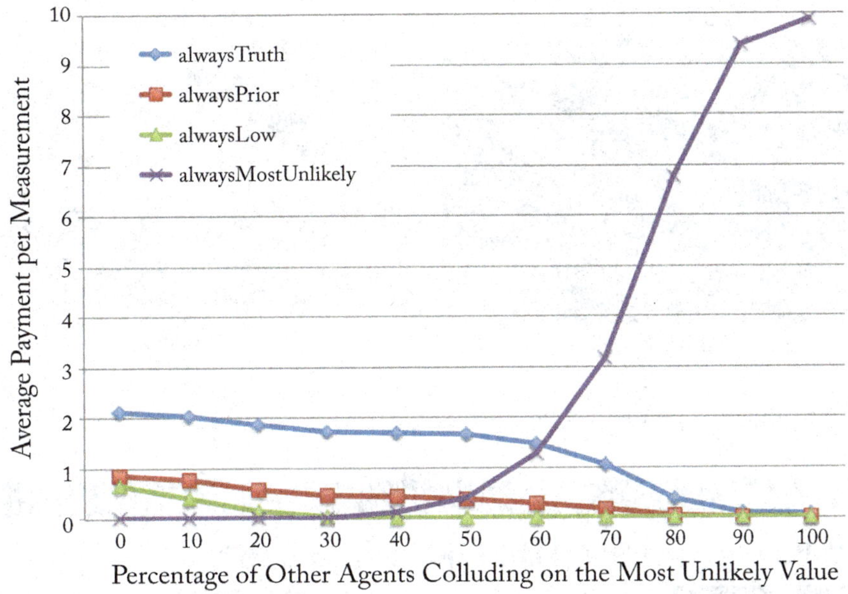

Figure 3.11: Average rewards for different collusive strategies, as a function of the fraction of agents adopting an uninformative strategy. Courtesy of Jason Li.

In Chapter 6, we describe the design and our experiences with the classical prediction market design in Swissnoise [55]. Here, we show how we implemented a novel peer prediction technique within this platform, that allowed us to compare the two methods.

One of the main drawbacks of classical prediction markets is that they require that the information that is provided can eventually be tied to a publicly verifiable outcome so that the securities can be paid off. This strongly limits the applicability of this technique: for example, it is not possible to elicit questions about hypothetical actions such as "What would be the success of a new bus line between stops X and Y?" or "What would be the outcome of a referendum on country X remaining in the EU?" when such actions are actually not planned. However, many of the questions that we would like to have predictions for are of this kind. One can use peer consistency to create prediction platforms that do not require that information is ever verified: rewards are given depending on consistency with other predictions.

Classical prediction markets are easy to understand because of the analogy with securities trading. The main challenge for adapting the peer consistency principle is to find a similar analogy (Figure 3.12). Swissnoise introduced the analogy of a *lottery* where agents can buy a ticket for a certain prediction outcome and day (Figure 3.13). At the end of the day, we run a lottery where all tickets bought that day are thrown into a pool. For each ticket, we randomly draw another ticket from the rest of the set, and apply the PTS mechanism to the two outcomes,

Figure 3.12: Analogies in prediction platforms.

Figure 3.13: Interaction for buying a lottery ticket on Swissnoise.

where the distribution R is taken from the prediction that was valid during that day. Following this, all tickets that were bought that day are aggregated with tickets of earlier days to form an updated predicted distribution of the different outcomes.

The game-theoretic analysis of peer consistency mechanisms assumes that agents are *risk-neutral*, i.e., they are indifferent between obtaining a very volatile reward and a completely sure award equal to the *expected* value of the volatile reward. However, Swissnoise is dealing with human agents, and they are known to be risk-averse. Thus, when this principle was first used on the Swissnoise platform, there was a bias against unlikely answers because their reward carries higher risk.

According to Fechner's law [23], people perceive variations as equivalent if they are *scale-invariant*, so that sensitivity to a stimulus is proportional to its *logarithm*. Applying this principle to risk-aversion would mean that rewards should increase *exponentially* with $1/R(x)$. This, however, proved too extreme for very unlikely values, so we used an average between $1/R$ and $e^{1/R}$.

To validate the accuracy that could be obtained through this reward scheme, for the same question the platform randomly assigned users to the classical prediction market and the peer prediction market version. For example, Figure 3.14 shows the price evolution for the two outcomes of the 2014 referendum on Scottish independence.

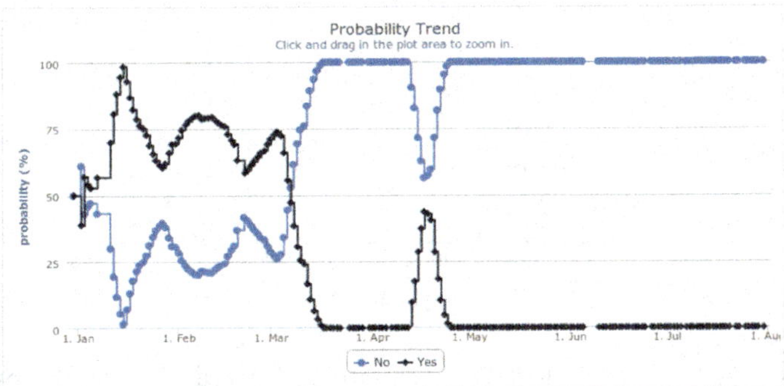

Figure 3.14: Price evolution for the question "Will Scotland be independent" using peer prediction.

It should be compared to the price evolution in the classical prediction market, shown in Figure 6.6, where wild price swings occured. They are due to two issues that plague classical prediction markets: it is very hard to tune the liquidity parameter (how much the price is influenced by share demand), and participants cashing in bets early changes predictions.

Both of these problems are absent in peer consistency, so not surprisingly the price evolution of the peer prediction market is a lot smoother and actually reflects the evolution of expectations very well.

Garcin and Faltings [55] report a comparison of the average accuracy of the two types of prediction markets on 32 different questions with good participation in both schemes. Figure 3.15 shows the accuracy, measured as the percentage of questions where the predicted probability of the correct answer exceeded a given probability threshold. Thus, for a question with 4 answers, if the threshold is 0.3, the question is counted as correct if the probability of the correct answer is predicted to be at least 0.3. We can see that both schemes obtain about the same accuracy; if we set the threshold at 0.5 the prediction is correct 72% of the time for the PTS version and 62% of the time for the classical prediction market.

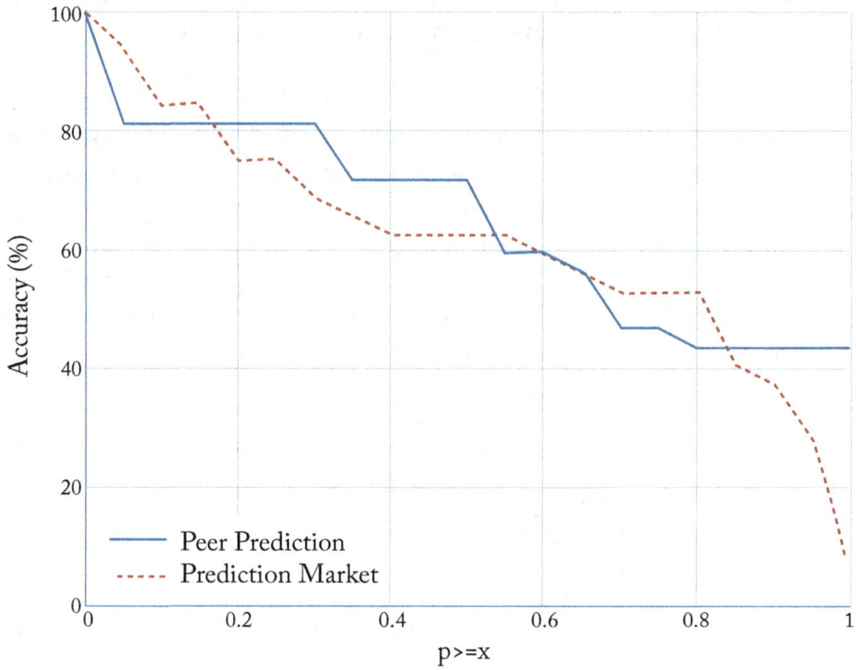

Figure 3.15: Accuracy comparison of the two versions of prediction platforms (based on [55]).

However, the peer prediction version generally has a more stable price evolution that is easier to interpret, and it was better appreciated by participants since rewards were instantaneous.[2] Finally, even though all questions on Swissnoise had a ground truth to allow comparison of the two schemes, the peer consistency version is much more widely applicable.

3.4.4 HUMAN COMPUTATION

Human computation platforms such as Amazon Turk connect requesters with a pool of workers who carry out small information tasks through the internet in return for small payments. A big issue is to ensure that workers actually put in the effort required to obtain the carry out the tasks that are requested of them. Requesters typically employ filtering mechanisms to identify unre-

[2] They, however, could not choose, but were randomly assigned to one or the other type.

liable workers, and give the same task to several workers to obtain a majority answer. However, these techniques fail to deal with the issue of bias.

Biases arise when workers carry out many similar tedious tasks and form a prior expectation of what the answer is likely to be. Workers will then develop a tendency to miss the unlikely answers. Bias can also arise from social influence and word-of-mouth [24]. If all workers develop the same bias, the redundancy obtained by using multiple workers cannot be used to eliminate the errors, and this is an important open problem in crowdsourcing.

However, it has been shown that using the peer truth serum as a worker incentive scheme can eliminate the answer bias [25]. It requires, however, that the bias is known, most naturally through the distribution of answers that have been given earlier. Faltings et al. [25] reported a study where the task was to count binoculars and cameras in an image, as shown in Figure 3.16. The actual number of such objects in the image is 34. The study tested conditions where the workers received no prior information, or where they were explicitly primed that the answer was likely to be around 34, or that it was likely to be around 60. The effect of priming was clearly visible: while without priming the answers were distributed normally around the correct answer of 34, and a similar distribution was obtained when priming to the correct value, when priming to 60 the distribution became skewed to larger values and the average was far off the true value. Figure 3.17 shows the influence of priming on the answers. Clearly priming has a strong effect and skews the answer distribution; the average error grows from 1.06 to 5.63.

Figure 3.16: Image used to test the effect of incentives on answer bias [26].

Four different incentive schemes were compared:

1. no incentive,
2. a vague bonus ("you receive a bonus if you do a good job"),
3. peer confirmation [27], which is equivalent to output agreement we discussed in Chapter 2, and

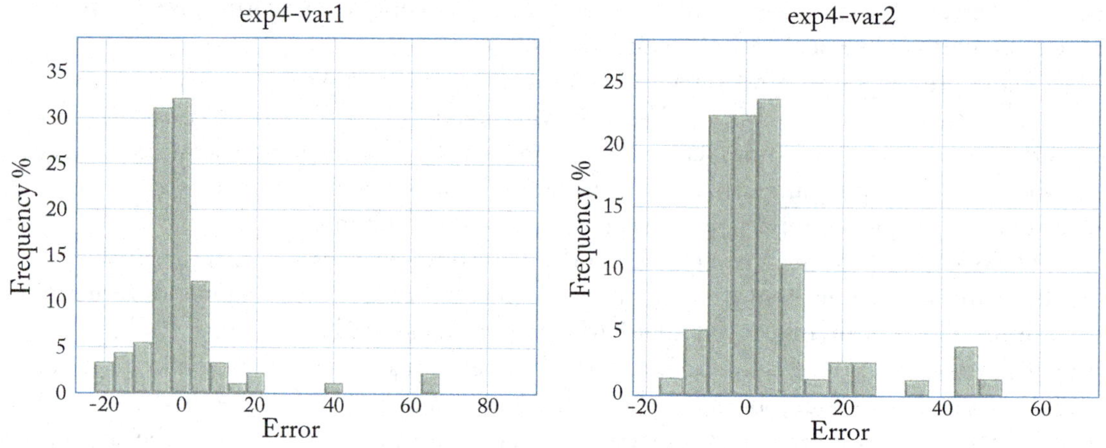

Figure 3.17: Distribution of answers without bonus and without priming (left), and with priming to a prior expection of 60 (right) (based on [25]).

4. the peer truth serum described above, using a discretized version of a distribution centered on the value used for priming.

Table 3.1 shows a comparison of different incentive schemes across the different priming scenarios. Clearly, we can see that while peer confirmation offers some help, only the peer truth serum is effective in eliminating the answer bias. We stress again that there is no alternative method for getting rid of such answer bias, and incentives are essential for obtaining high-quality data.

Table 3.1: Comparison of different incentive schemes in the presence of priming

Bonus Scheme	Priming	Average Error	t-test
none	60	5.6316	
vague	60	6.6563	p = 0.3782
peer conf.	60	3.3429	p = 0.1306
PTS	60	0.8000	p = 0.0088
none	34	2.9434	
vague	34	9.0984	p = 0.0110
peer conf.	34	2.4194	p = 0.4020
PTS	34	2.1667	p = 0.3731

CHAPTER 4

Nonparametric Mechanisms: Multiple Reports

Knowing agent beliefs As we have seen, incentives for encouraging agents to be truthful hinge crucially on their *beliefs* and the way they are affected by observations. Thus, the mechanisms we have seen in Chapter 3 all have some characterization of agent beliefs as a parameter.

This is not desirable for two reasons. The first is that it is very hard for the center to guess these parameters correctly. The second is that as the same mechanism is applied uniformly to a large population of agents, their beliefs have to be very uniform.

Thus, it would be desirable to have mechanisms that do not require these beliefs to be known. In this and the next chapter, we are going to see mechanisms that do this either by

- eliciting beliefs from the reporting agents through additional reports (this chapter), or by
- learning the necessary probability distributions through observation of the data reported by the agents (Chapter 5).

4.1 BAYESIAN TRUTH SERUM

The idea of the *Bayesian Truth Serum* [28] is to ask agents to provide two reports: an *information report* x_i that contains the data of interest, and a *prediction report* F_i that contains a prediction of what other agents report as data, which is exactly the agent's posterior belief q about the data.

Both reports receive a *score* and the reward given to the agent is obtained as the sum of both scores:

$$\tau_{BTS}(x_i, F_i, \ldots) = \underbrace{\tau_{info}(x_i, \ldots)}_{\textit{information score}} + \underbrace{\tau_{pred}(F_i, \ldots)}_{\textit{prediction score}}.$$

The original Bayesian Truth Serum [28] computes the scores from the following two quantities:

- *freq*(x)—(normalized) frequency of reports equal to x:

$$freq(x) = \frac{num(x)}{n}.$$

Mechanism 4.1 The original Bayesian truth serum (BTS).

1. Center gives the same task to agents $A = \{a_1, \ldots, a_k\}$. Each agent a_i reports an *information report* x_i and a *prediction report* F_i, where the prediction report is an estimate of the distribution of answers x_i among agents in A.
2. To compute the score of agent a_i, the center computes the histogram of information reports $freq_{-i}(x)$ and the geometric mean of prediction reports $gm_{-i}(F)$, where the reports of a_i are excluded from the mean.
3. The center computes the prediction score $\tau_{pred} = -D_{KL}(freq_{-i}(x)||F_i(x))$ and the information score $\tau_{inf} = \ln freq_{-i}(x_i) - \ln gm_{-i}(x_i)$.
4. The center rewards a_i with a payment proportional to:

$$\tau_{BTS}(x_i, F_i) = \tau_{inf} + \tau_{pred}.$$

- *gm*—geometric mean of agents' predictions F_j:

$$\log gm(x) = \frac{1}{n} \sum_j \log f_j(x).$$

The prediction score $\tau_{pred}(F_i, \ldots)$ is obtained by evaluating the prediction report F_i over all peers using the logarithmic scoring rule:

$$\tau_{pred}(F_i, \ldots) = \frac{1}{n} \sum_j \log(f_i(x_j)) + C = \sum_x freq(x) \cdot \log(f_i(x)) + C.$$

By setting the constant $C = -\sum_x freq(x) \cdot \log(freq(x))$ we obtain:

$$\tau_{pred}(F_i, \ldots) = \sum_x freq(x) \cdot \log \frac{f_i(x)}{freq(x)} = -KL(freq(x)||F_i(x))$$

so that the prediction score $\tau_{pred}(F_i, \ldots)$ measures how bad F_i is compared to the actual frequencies observed in the information reports. This provides the incentive for an agent to be as truthful as possible about the prediction report.

The principle for scoring the information report is to compute how much *better* the information is compared to what would be obtained by using the geometric mean *gm* of agents' prediction reports. For this we use the logarithmic scoring rule to compare:

- the score of the observed frequency of the reported value x_i:

$$C - \log freq(x_i)$$

- the average score of the prediction reports of all agents:

$$\frac{1}{n}\sum_{j=1}^{n}\left[C - \log f_j(x_i)\right] = C - \log gm(x_i)$$

to obtain for the information score:

$$\tau_{info}(x_i, \ldots) = \log \frac{freq(x_i)}{gm(x_i)}.$$

The resulting mechanism is summarized in Mechanism 4.1.

What is the incentive given by the information score? In a truthful equilibrium, we can consider the term $freq(x_i)$ to be the actual probability that the correct answer is x_i, and $\log freq(x_i)$ is thus maximized for reporting the most likely value. To normalize the score for reporting according to prior knowledge to zero, we subtract the average score that would be obtained by reporting according to the prior, which is $gm(x_i)$. If we take $gm(x_i)$ to be the prior probability P and $freq(x_i)$ to be the posterior probability Q, the expected information score using the original BTS mechanism would be:

$$E[\tau_{info}] = \sum_i q(x_i) \log \frac{q(x_i)}{p(x_i)} = D_{KL}(Q||P),$$

thus the Kullback-Leibler divergence between the prior and posterior distributions.[1]

The following properties are shown in Prelec [28].

Theorem 4.1 *Given a large enough population of agents, the original Bayesian Truth Serum has cooperative strategy as a strict Bayes-Nash equilibrium.*

Thus, in order to achieve truthfulness with BTS, it is important to have a large enough number of agents. In fact, the lower bound on the number of needed agents depends on agents beliefs. We will see in the following subsections how to relax the requirement for a large population.

As an additional observation, Prelec [28] notes that when given equal weight, the sum of all BTS scores adds up to zero; the sum of prediction scores:

$$\begin{aligned}
\sum_i \tau_{pred}(F_i, \ldots) &= \sum_i \sum_x freq(x) \cdot \log \frac{f_i(x)}{freq(x)} \\
&= \sum_x \sum_i \frac{num(x)}{n} \cdot [\log f_i(x) - \log freq(x)] \\
&= \sum_x num(x) \cdot \log gm(x) - \sum_x num(x) \cdot \log freq(x) \\
&= \sum_i \log \frac{gm(x_i)}{freq(x_i)} =_{def} -\sum_i \tau_{info}(x_i, \ldots)
\end{aligned}$$

[1]The exact reasoning behind BTS is more complex, but here we give an intuitive analogy to the mechanisms discussed in this book.

is just the negative of the sum of information scores.

This zero-sum structure implies the collusive strategies involving all agents are not profitable, since they cannot increase the total revenue that the center pays to all of them. Thus, we do not have to worry about uninformative equilibria where everyone reports the same value. On the other hand, this also means that the mechanism does not reward agents for increased accuracy: the sum of payments remains the same, whether agents work hard or not. In fact, since the game is a contest among agents, it could be that even agents that put in a lot of effort get negative rewards.

It would be appropriate for example for assigning quality scores to the information provided by the agents. Note that we can easily make the expected reward positive by giving the information report a higher weight relative to the prediction report.

4.2 ROBUST BAYESIAN TRUTH SERUM

A major problem of the BTS mechanism is that the distributions observed on the reported data can be quite far from the true probability distributions. If there are only n reports in total, the frequency of any value is a multiple of $1/n$, and given the sampling there is a good chance that it is quite far from the true probability. This can significantly disturb the incentives of the original BTS mechanism.

Mechanism 4.2 The robust Bayesian truth serum (BTS).

1. Center gives the same task to agents $A = \{a_1, \ldots, a_k\}$. Each agent a_i reports an *information report* x_i and a *prediction report* F_i, where the prediction report is an estimate of the distribution of answers x_i among agents in A.

2. The center picks a random peer agent $a_j \in A$ and computes the reward to agent a_i as

$$\tau_{decomp}(x_i, F_i, x_j, F_j) = \underbrace{\frac{\mathbf{1}_{x_i=x_j}}{f_j(x_i)}}_{\text{information score}} + \underbrace{f_i(x_j) - \frac{1}{2}\sum_z f_i(z)^2}_{\text{prediction score}}.$$

Therefore, *robust* versions of BTS have been developed that work even with a small number of reports [33, 34]. They keep the decomposable structure of the score into an information score and a prediction score, where the information score gives an incentive for truthfulness based on another agent's prediction report, and the prediction score uses a proper scoring rule against the information report. For example [34]:

$$\tau_{decomp}(x_i, F_i, x_j, F_j) = \underbrace{\frac{\mathbf{1}_{x_i=x_j}}{f_j(x_i)}}_{\textit{information score}} + \underbrace{f_i(x_j) - \frac{1}{2}\sum_z f_i(z)^2}_{\textit{prediction score}}$$

uses the PTS mechanism to score the information report, which requires the self-predicting condition to hold, as in the original BTS mechanism.

As an example to illustrate robust decomposable BTS mechanisms, consider eliciting a variable with 3 values 0, 1 and 2, and the agent prior and posterior beliefs shown in Table 4.1. Suppose that agent A_i observes the value $o = 0$ and that its peer agent A_j is honest. The prediction score is computed using the quadratic scoring rule $QSR(A, x) = a(x) - \frac{1}{2}\sum_z a(z)^2$. Thus, the prediction score that A_i expects for its prediction report $F_i = q_0$ is:

Table 4.1: Agent prior and posterior beliefs for eliciting a variable with three values

o	0	1	2
$p(o)$	0.1	0.5	0.4
$q_0(o)$	0.3	0.2	0.2
$q_1(o)$	0.4	0.6	0.3
$q_2(o)$	0.3	0.2	0.5

$$\begin{aligned} E(\tau_{pred}(F_i, \ldots)) &= q_0(0) \cdot QSR(F_i, 0) + \\ &+ q_0(1) \cdot QSR(F_i, 1) + q_0(2) \cdot QSR(F_i, 2) = \\ &= 0.3 \cdot 0.13 + 0.4 \cdot 0.23 + 0.3 \cdot 0.13 = 0.17. \end{aligned}$$

On the other hand, if A_i provided an inaccurate prediction report of $F_i = (f_i(0), f_i(1), f_i(2)) = (0.5, 0.2, 0.3)$, it will believe its prediction score to be lower:

$$\begin{aligned} E(\tau_{pred}(F_i, \ldots)) &= q_0(0) \cdot QSR(F_i, 0) + \\ &+ q_0(1) \cdot QSR(F_i, 1) + q_0(2) \cdot QSR(F_i, 2) = \\ &= 0.3 \cdot 0.31 + 0.4 \cdot 0.01 + 0.3 \cdot 0.11 = 0.13. \end{aligned}$$

In general, because of the proper scoring rule, we can show that $E(\tau_{pred}(F_{honest}, \ldots)) > E(\tau_{pred}(F_{dishonest}, \ldots))$.

Now consider the information score. If A_i honestly reports its observation 0, it will expect to obtain a score of:

$$\begin{aligned} E(\tau_{info}(x_i = 0, \ldots)) &= E\left(\frac{\mathbf{1}_{x_j=0}}{f_j(0)}\right) = \frac{q_0(0)}{f_j(0)} \\ &= \frac{q_0(0)}{q_0(0)} = 1 \end{aligned}$$

whereas for an incorrect report it expects a lower score, for example when reporting a value of 1 (while having observed 0):

$$E(\tau_{info}(x_i = 1, \ldots)) = E\left(\frac{\mathbf{1}_{x_j=1}}{f_j(1)}\right) = \frac{q_0(1)}{f_j(1)}$$
$$= \frac{q_0(1)}{q_1(1)} = 0.67$$

and, in general, we can show (see above) that $E(\tau_{info}(x_{honest}, \ldots)) > E(\tau_{info}(x_{dishonest}, \ldots))$.

For robust BTS, we can show the following property [34].

Theorem 4.2 *Provided that agent belief updates satisfy the self-predicting condition, the robust BTS mechanism (Mechanism 4.2) has cooperative strategy as a strict Bayes-Nash equilibrium.*

Decomposable mechanisms It is interesting to note that both the original BTS score and RBTS score the information report separately from the prediction report. We call such mechanisms *decomposable*, as an agent's information score is independent of its prediction report, while an agent's prediction score is independent of its prediction report. Typically, the theoretical analysis of such mechanisms is relatively simple, while their structure often allows intuitive interpretations of scores (for example, see Radanovic [36]). From the game-theoretic point of view, however, the class of decomposable mechanisms is not complete in a sense that one cannot achieve truthful elicitation without assuming additional constraints on a common belief system. More formally, it is shown in Radanovic and Faltings [34].

Theorem 4.3 *There does not exist a decomposable BTS mechanism that has cooperative strategy as a strict Bayes-Nash equilibrium for a general structure of agents' beliefs.*

As in the case of Theorem 3.3, Theorem 4.3 holds for any agent beliefs that are statistically relevant, which makes it quite broadly applicable. In order to achieve truthful elicitation for a general case of an agents' common belief system, we consider in the next section a non-decomposable type of BTS mechanism.

4.3 DIVERGENCE-BASED BTS

The Bayesian Truth Serum, as we have seen above, still has the drawback that it either requires a large number of agents, or (for the robust versions) requires the self-predicting constraint to hold.

We are now going to show an alternative approach where we penalize agents for inconsistencies in their reports, developed independently by Radanovic and Faltings [35] and Kong and Schoenebeck [37]. We will keep the same principle for the prediction score, where we score the report against a peer report using a proper scoring rule. However, the information score

Mechanism 4.3 The divergence-based Bayesian truth serum (DBTS).

1. Center gives the same task to agents $A = \{a_1, \ldots, a_k\}$. Each agent a_i reports an *information report* x_i and a *prediction report* F_i, where the prediction report is an estimate of the distribution of answers x_i among agents in A.

2. The center picks a random peer agent $a_j \in A$ and computes the reward to agent a_i as:

$$pay(x_i, F_i, x_j, F_j) = \underbrace{-\mathbf{1}_{x_i = x_j \wedge D_{KL}(F_i || F_j) > \Theta}}_{\text{information score}} + \underbrace{f_i(x_j) - \frac{1}{2}\sum_z f_i(z)^2}_{\text{prediction score}}.$$

will *penalize* inconsistencies, in particular agents who give the same information reports but significantly different prediction scores.

We define the *divergence-based BTS* by the following score functions:

- a prediction score as in RBTS, using the quadratic scoring rule:

$$pay_{pred}(x_i, F_i, x_j, F_j) = f_i(x_j) - \frac{1}{2}\sum_z f_i(z)^2; \quad \text{and}$$

- an information score that penalizes absence of a divergence $> \Theta$ for prediction reports of agents with different information reports:

$$pay_{info}(x_i, F_i, x_j, F_j) = \begin{cases} -1 & \text{if } x_i = x_j \wedge D(F_i || F_j) > \Theta \\ 0 & \text{otherwise} \end{cases}$$

$$pay(x_i, F_i, x_j, F_j) = \underbrace{-\mathbf{1}_{x_i = x_j \wedge D_{KL}(F_i || F_j) > \Theta}}_{\text{information score}} + \underbrace{f_i(x_j) - \frac{1}{2}\sum_z f_i(z)^2}_{\text{prediction score}}.$$

The information score still requires a parameter Θ that would need to set correctly by the center. We can eliminate the need for this parameters by comparing divergence with a randomly selected third agent who reported a different value (and thus should have a different posterior distribution):

$$pay_{info}(\ldots) = \begin{cases} -1 & \text{if } x_i = x_j \neq x_k \wedge D(F_i || F_j) > D(F_i || F_k) \\ 0 & \text{otherwise.} \end{cases}$$

Consider the divergence-based BTS mechanism on the same example we used above (see Table 4.1). The prediction score is the same as for the robust BTS mechanism given above. Consider first the parametric version with $\Theta = 0.01$, and let agents honestly report their posterior beliefs as prediction reports. The information score is then:

$$\begin{aligned} E(\tau_{info}(x_i = 0, \ldots)) &= -q_0(0) \cdot \mathbf{1}_{D(q_0 \| q_0) > 0.01} \\ &= 0. \end{aligned}$$

On the other hand, if it reports incorrectly the value 1, it expects an information score:

$$\begin{aligned} E(\tau_{info}(x_i = 0, \ldots)) &= -q_0(1) \cdot \mathbf{1}_{D(q_0 \| q_1) > \Theta} \\ &= -0.4 \cdot \mathbf{1}_{(0.2-0.3)^2 + (0.6-0.4)^2 + (0.2-0.3)^2 > 0.01} = -0.4 \end{aligned}$$

and we can again see that $E(\tau_{info}(x_{honest}, \ldots)) > E(\tau_{info}(x_{dishonest}, \ldots))$.

Under the assumption of a technical condition on agents' beliefs that is entailed by either the conditions of the original BTS or the conditions of the robust BTS mechanism, it is possible to show [35] the following.

Theorem 4.4 *The divergence-based BTS has truthful reporting as a strict Bayes-Nash equilibrium.*

A similar divergence-based mechanism is introduced in Kong and Schoenebeck [37], but still with the assumption of a common prior belief that is not known to the mechanism. In Kong and Schoenebeck [4], the same authors provide an extended framework for such mechanisms that include not only divergence but also other types of comparisons applied to prediction reports, in particular mutual information and information gain. Their work pays particular attention to ruling out equilibria where agents do not provide true information.

Continuous values The divergence-based BTS mechanism has some important advantages. It works even for small populations of agents, and it does not require agents to have identical prior beliefs. This possibility has also allowed to extend it to reports of continuous values [35] that occur particularly in sensor data. We will now describe a parametric version of the divergence-based BTS designed for continuous domain, which is transformable to a nonparametric mechanisms for certain types of agents' beliefs (see Radanovic and Faltings [35] for more details). Notice that the parametric mechanism we are about to describe has a very weak requirements on how to select a proper value of its parameter. Thus, although the mechanism is technically parametric, it is righteously described in this chapter.

Clearly, the direct application of the divergence BTS mechanism is not possible since the penalty score is not well defined for continuous values. To define a penalty score that compares information reports in a more meaningful way, we discretize real domain into intervals of equal sizes, where the size is chosen randomly, while the starting point of the discretization procedure is defined by the value of the agent's information report. Then, the information reports of the agent and her peer are considered to be similar if they fall into the same interval. This full transformation of the divergence-based BTS can be described by the following steps.

1. Each agent i is asked to provide the information report x_i and the prediction report F_i, as in the divergence-based BTS.

2. For each agent i, the mechanism samples a number δ_i from a uniform distribution, i.e., $\delta_i = rand((0, 1))$.[2] The continuous answer space is then uniformly discretized with the discretization interval of a size δ_i and the constraint that value x_i is in the middle of the interval it belongs to. We denote the interval of a value x_i by Δ_x^i. The constraint can then be written as $x_i = \frac{\max \Delta_x^i - \min \Delta_x^i}{2}$.

3. Finally, an agent i is scored using a modified version of the divergence-based BTS score:[3]

$$\underbrace{-\mathbf{1}_{x_j \in \Delta_x^i \wedge D_{KL}(F_i||F_j) > \delta_i} \cdot \Theta}_{\textit{information score}} + \underbrace{\log(f_i(x_j))}_{\textit{prediction score}}. \tag{4.1}$$

Notice that the only restriction for the properness of the above mechanism Θ is large enough. However, there is a tradeoff between the value of Θ and the expected value of margin difference of the information score between truthful and non-truthful reporting. That is, the larger Θ is, the smaller the expected punishment is for an agent who deviates from truthful reporting. Moreover, now the prediction report is a probability density function. In the case of parametric distribution functions that are often used in practice, reporting predictions comes down to reporting relatively few small number of real valued parameters.

Under a monotonicity condition about the divergence of prediction reports and information reports, it is possible to show (see [35]):

Theorem 4.5 *The divergence-based BTS for continuous signals (Mechanism 4.4) has cooperative strategy as a strict Bayes-Nash equilibrium.*

Generality of divergence-based BTS We have seen that the divergence-based BTS elicits truthful responses under fairly general conditions. For data with continuous values, it is required that agents share a common belief system. One might wonder if it is possible to relax this requirement, especially since for discrete values the divergence-based BTS *does* allow deviation from this condition. Unfortunately, this is not achievable in the BTS setting [35, 36]. For a very natural class of belief systems that are based on Gaussian distributions, even small deviations from the common belief system condition are hard to deal with:

Theorem 4.6 *There exists a parametric class of belief systems such that no BTS mechanism simultaneously has:*

[2]We take the discretization interval to be of a size 1, but one can make it larger or smaller. Furthermore, δ_i can be sampled from a different type of distribution, as long as it has the full support over the discretization interval.

[3]For simplicity, we used logarithmic scoring rule for the continuous version of the divergence-based BTS. One can apply the quadratic scoring rule as well (as in the divergence-based BTS), but applied to elicitation of probability density functions.

Mechanism 4.4 The divergence-based Bayesian truth serum (DBTS) for continuous values.

1. Center gives the same task to agents $A = \{a_1, \dots, a_k\}$. Each agent a_i reports an *information report* x_i and a *prediction report* F_i, where the prediction report is an estimate of the distribution of answers x_i among agents in A.
2. For each agent a_i, the center uniformly discretizes the continuous answer space with the discretization interval of a size δ_i and the constraint that value x_i is in the middle of the interval it belongs to.
3. The center picks a random peer agent $a_j \in A$ and computes the reward to agent a_i as:

$$pay(x_i, F_i, x_j, F_j) = \underbrace{-\mathbf{1}_{x_j \in \Delta_x^i \wedge D(F_i || F_j) > \delta_i \cdot \Theta}}_{\textit{information score}} + \underbrace{\log(f_i(x_j))}_{\textit{prediction score}}.$$

- *cooperative strategy as a strict Bayes-Nash equilibrium when all agents have a common belief system B; and*
- *cooperative strategy as a strict Bayes-Nash equilibrium when one agent has a belief system $\hat{B} \neq B$ and all the other agents have a common belief system B.*

In a sense, this result implies the generality of the divergence-based BTS in the standard BTS settings. Furthermore, it demonstrates the difficulty of eliciting continuous private information in non-parametric class of peer consistency mechanisms.

Practical concerns All of the BTS mechanisms require reporting agents to also form an explicit opinion about what other agents may report. This data can be significantly more complex than the information report itself. Furthermore, for the incentive properties to hold, the agent must not know about the reports that have been received by other agents—otherwise it would be easy to submit a prediction report that exactly matches this distribution. They are thus unsuitable for applications such as opinion polls or reputation forums that publish this information continuously.

4.4 TWO-STAGE MECHANISMS

Several works in the literature have proposed mechanisms where agents two reports at different times. They offer interesting advantages if the setting does allow for eliciting information in two separate stages.

Witkowski and Parkes [38] propose a mechanism where agents first provide a prediction report before observing the phenomenon, followed by another report after observing the phenomenon. The change in the two reports can be used to derive the actual observation. Zhang and Chen [39] propose a mechanism where agents first report their information reports, and then form a prediction report while taking into account the information report given by a peer agent. They show that this dependency can be exploited to ensure truthfulness for a slightly more general class of belief structures than the RBTS mechanism described above.

4.5 APPLICATIONS

Due to the complexity of the prediction report, the Bayesian Truth Serum has not been widely experimented so far. An experiment reported in Prelec and Seung [29] elicited predictions about the state capitals in the U.S. As these are often not the biggest and best-known cities, there is a lot of confusion, and in fact it turned out that in their experiment it was often the case that the majority gave the wrong answer. For example, for the state of Illinois, most students wrongly believed the capital to be Chicago.

The experiment asked 51 students at MIT and 32 students at Princeton University a set of 50 questions, asking for each state whether the most populous city was the capital of that state. Thus, for example, it would ask: "Is Chicago the capital of Illinois?".

Students showed poor knowledge of geography and only answered slightly better than chance, with the average student getting 29.5 correct answers at MIT and 31 at Princeton. The majority decisions were slightly better, giving correct answers for 31 questions at MIT and 36 questions (with 4 ties) at Princeton.

The subjects showed a strong correlation between prediction and information reports: those who believed the answer to a question to be YES believed that on average 70.3% of the others would also answer YES, while those who answered NO believed on average only 49.8% would answer YES.

To evaluate the BTS mechanism, the researchers considered a voting scheme where the answers were weighted by the BTS score computed for the answers of the respondent, and the chosen answer was the one with the highest sum of BTS scores. They reported a dramatic effect on the accuracy: in the MIT sample, the number of correct majority decisions rises from 31–41, and in the Princeton sample, it rises from 36–42 (still with 4 ties).

The experiment also showed a strong correlation between the number of correct answers and the BTS score obtained by respondents.

Weaver and Prelec [30] report another study where BTS is shown to provide effective incentives against the overclaiming effect. In this experiment, human subjects are asked whether they recognize real and fake brand names. Giving a bonus for every recognized name leads to overclaiming, which can be measured by the fraction of fake brand names that are supposedly recognized. The study shows that BTS is effective in countering this bias.

John et al. [31] show a similar effect, where psychologists underclaim their use of questionable research practices—here underclaiming is caused not by monetary rewards, but by embarassment that is countered (but probably not entirely compensated) by BTS.

BTS is also proposed as a scoring scheme for better aggregating information according to confidence. Prelec et al. [32] report on four different studies, including the one of state capitols described above, where the BTS score of answers is used as a weighting criterion for aggregating information, rather than an incentive for obtaining accurate information in the first place.

CHAPTER 5

Nonparametric Mechanisms: Multiple Tasks

Another strategy for obtaining nonparametric mechanisms is to *learn* the parameters from the data submitted by the agents themselves, during some fixed time interval. This works whenever we have agents providing data about multiple (and ideally many) very similar phenomena within a short time interval, as shown in Figure 5.1.

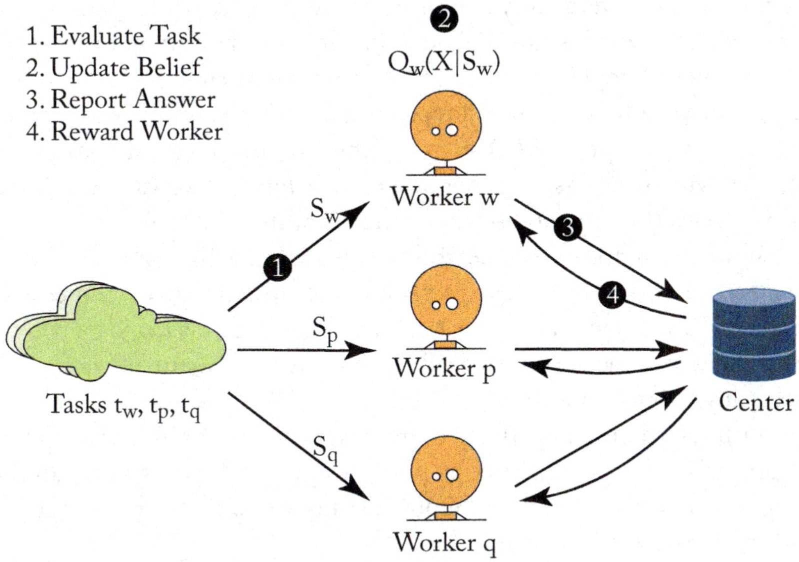

Figure 5.1: Scenario for multi-task mechanisms.

5.1 CORRELATED AGREEMENT

For example, we can observe the frequency of different submitted reports, and use this information to make the expected reward of any random reporting strategy equal to zero. Dasgupta and Ghosh [41] introduced such a mechanism for binary-valued queries. It applies peer consistency with a constant reward for each value, and subtracts the probability $r(x)$ that a randomly chosen report would also match the same answer a, which is just the probability that a random answer is equal to x and can be approximated from the observed data. In this way, the expected reward for reporting randomly according to *any* distribution (including only reporting a single value) is exactly equal to zero.

One way to obtain $r(x)$ would be to just take the frequency of x in the data. However, the mechanism in Dasgupta and Ghosh [41] choses a more elegant solution. It randomly selects a report for an unrelated task w, and creates a random variable that is equal to 1 when $w = a$ and 0 otherwise. The expected value of this proxy variable is just equal to $r(a)$. The resulting payment rule is thus:

$$pay(x, y) = \underbrace{\mathbf{1}_{x=y}}_{\text{output agreement}} - \underbrace{\mathbf{1}_{x=w}}_{\text{proxy for r}},$$

where w is a random peer answer to a *different* task.

This mechanism has several very useful properties. First, while it admits uninformative equilibria, their expected reward is exactly equal to zero, and they are not interesting for agents. Second, it is very simple to implement, and requires no parameters.

The mechanism incentivizes truthful reporting of an observation x as long as its posterior probability increases over the prior: $q(x) \geq p(x)$. Thus, it encourages truthful reporting as long as x only correlates with itself, as all other values will have a negative expected reward. For two values, this is always the case, and so no further condition is necessary.

However, if we want to generalize this mechanism in the obvious way to more than two values, we need to ensure that there is no correlation among the values—any value that is positively correlated with x will also give a positive expected reward.

Can we generalize a mechanism based on output agreement to more than two values without imposing any stricter conditions?

Consider a value z that is negatively correlated with x. For an agent that observes x, it would never be profitable to report z, since its posterior probability is lower than the prior and so it's expected reward is negative. However, reporting any value that is positively correlated with x results in a positive expected payment.

The *correlated agreement* mechanism [42] generalizes this idea to scenarios with more than two values, under the following assumptions.

- Agents answer multiple tasks and use the same strategy everywhere.
- Agents and center know and agree on sign of correlation among each answer pair for different agents/same task.
- Nothing else is known about agent beliefs.
- Distinguishing correlated values is not important.

Mechanism 5.1 The correlated agreement mechanism.

1. Center gives a set of similar tasks $\mathcal{T} = \{t_1, \ldots, t_m\}$ to agents $A = \{a_1, \ldots, a_k\}$ such that every agent a_i solves multiple tasks and every task is solved by multiple agents; a_i reports data x_i.
2. Center computes the matrix of correlations $\mathbf{\Delta}$ on the signal distributions $\Pr(s)$ expected for the tasks such that $\Delta(x, y) = \Pr(x, y) - \Pr(x)\Pr(y)$. Alternatively, the correlations can be computed on the collection of answers received from the agents. It derives the score matrix $S(x, y)$ where $S(x, y) = 1$ if $\Delta(x, y) > 0$ and $S(x, y) = 0$ otherwise.
3. To compute the reward to a_i for its answer x_i to task t_m, the center randomly selects a *peer agent* a_j that has also submitted an answer x_j for t_m, and let y_i and y_j be two answers for other tasks submitted by a_i and a_j.
4. Center pays agent a_i a reward proportional to:

$$\tau(x_i, x_j, y_i, y_j) = S(x_i, x_j) - S(y_i, y_j).$$

To thwart the incentive to report correlated values, the correlated agreement mechanism gives a constant reward whenever report x_i is *positively correlated* with a randomly chosen peer's answer x_j. More specifically, we define the matrix Δ of value correlations:

$$\Delta(x, y) = \Pr(x, y) - \Pr(x)\Pr(y)$$

and define the *score* for agent report x and peer report y as:

$$S(x, y) = \begin{cases} 1 & \text{if } \Delta(x, y) > 0 \\ 0 & \text{otherwise.} \end{cases}$$

To discourage random reporting, it compares the score $S(x_i, x_j)$ for *same* task t_1 with the score for randomly chosen *different* tasks using reports y_i of agent i for t_2 and y_j of peer agent j for t_3 to obtain the payment:

$$\tau(x_i, x_j, y_i, y_j) = S(x_i, x_j) - S(y_i, y_j).$$

We can see that using this mechanism, truthful reporting of the signal is the best strategy by considering that the expected payment for truthful reporting is the sum of all positive entries in Δ:

$$E[pay] = \sum_{i,j} \Delta(x_i, x_j) S(x_i, x_j) = \sum_{i,j,\Delta(x_i,x_j)>0} \Delta(x_i, x_j).$$

Non-truthful strategies would sum different elements, and some of them would not be positive, so it can only achieve a smaller sum. Thus, truthful strategies result in the highest-paying equilibrium! We note that Kong and Schoenebeck [4] provide an alternative proof using an information-theoretic framework.

Clearly, the incentives for truthfulness also serve to incentivize the effort required to find out what the correct observation is. However, as the scheme gives the same payoff for reporting the true value and any positively correlated value, it cannot distinguish among these values. On one hand, this can be positive in situtations where there are strongly correlated values and agent belief updates may not even satisfy the self-predicting condition (Definition 1.4)—the CA mechanism would not provide incentives to report an incorrect value. On the other hand, when correlations are weaker, the CA mechanism cannot be used to obtain information with arbitrary precision.

Consider how this mechanism will work on the example of airline service we introduced in Chapter 3. Assume the following joint probability distribution for an agent and a randomly chosen peer:

		peer experience	
		b	g
agent	b	0.06	0.05
experience	g	0.05	0.84

so that probabilities of airline service are $(\Pr(good) = 0.89, \Pr(bad) = 0.11)$. The corresponding Δ-matrix is:

$$\Delta = \begin{array}{c|cc|} & b & g \\ \hline b & \underbrace{0.06 - 0.11^2}_{=0.0479} & \underbrace{0.05 - 0.11 \cdot 0.89}_{=-0.0479} \\ g & \underbrace{0.05 - 0.89 \cdot 0.11}_{=-0.0479} & \underbrace{0.84 - 0.89^2}_{=0.0479} \\ \hline \end{array}$$

Consider two example strategies that an agent might adopt over multiple reports: always report truthfully, or always report good service. They result in different expected scores for unrelated tasks that become the negative term in the payment function:

1. always truthful: probability of matching on unrelated tasks = $0.84^2 + 0.06^2 = 0.709$; or
2. always report good service: probability of matching on unrelated tasks = 0.84.

Consider now an agent that observed bad service, and adopts $\hat{q}_b(g) = 0.45$. Depending on whether it reports "bad" or "good" (first or second strategy), it can expect a reward of:

$$1 : E[pay(\text{``}bad\text{''})] = 0.55 \cdot \underbrace{Pay(b,b)}_{=1} + 0.45 \cdot \underbrace{Pay(b,g)}_{=0} - 0.709 \quad = \quad 0.55 - 0.709 = -0.159$$

$$2 : E[pay(\text{``}good\text{''})] = 0.55 \cdot \underbrace{Pay(g,b)}_{=0} + 0.45 \cdot \underbrace{Pay(g,g)}_{=1} - 0.84 \quad = \quad 0.45 - 0.84 = -0.39.$$

Clearly, the truthful strategy results in a lower loss. Let's also consider what happens when the agent observed good service, and adopts $\hat{q}_b(g) = 0.94$. Its expected payments are:

$$1 : E[pay(\text{``}good\text{''})] = 0.06 \cdot \underbrace{Pay(g, b)}_{=0} + 0.94 \cdot \underbrace{Pay(g, g)}_{=1} - 0.709 = 0.94 - 0.709 = 0.231$$
$$2 : E[pay(\text{``}good\text{''})] = 0.06 \cdot \underbrace{Pay(g, b)}_{=0} + 0.94 \cdot \underbrace{Pay(g, g)}_{=1} - 0.84 = 0.94 - 0.84 = 0.1.$$

So that, overall, the expected payoffs for the two strategies are (given that we have 15% bad and 85% good service:

$$1 : 0.11 \cdot (-0.159) + 0.89 \cdot 0.231 = 0.1881$$
$$2 : 0.11 \cdot (-0.39) + 0.89 \cdot 0.1 = 0.0461.$$

Thus, the correlated agreement mechanism works well for this example, and provides quite an efficient separation between truthful and non-truthful strategies.

For the CA mechanism we presented here, it is possible to show [42]:

Theorem 5.1 *The CA mechanism is maximally strong truthful among all multi-task mechanisms that only use knowledge of the correlation structure of signals.*

Here, strong truthfulness refers to a property in which truthfulness (cooperativeness) is a strict Bayes-Nash equilibrium whose payoff is greater than that of all other strategy profiles. The mechanism can also attain ex-post subjective equilibrium if the correlation structure of signals is agent specific, which can be achieved by learning it from elicited data.

An important limitation of the CA algorithm is that the correlations expressed by the Δ matrix must be the same for all agents. In reality, each pair of agents could have different correlations, depending on how similar they are in judging their observations. Modeling this situation completely for n agents would require learning $n(n-1)/2$ Δ-matrices, which in practice is intractable. A solution for this is to cluster the agents into groups that share similar judgement, either by the Δ-matrices or by the their confusion matrix according to the model of Dawid and Skene [5] that we introduced in Section 1.3.

Agarwal et al. [43] show how clusters can be learned from reports submitted by information agents without compromising the incentive properties of the CA scheme. The clustering induces two kinds of errors, a model error ϵ_1 for approximating the individual relations by clusters, and a sample error ϵ_2 for imperfectly learning the clusters from a finite number of samples. The algorithm presented in Agarwal et al. [43] results in truthtelling as an $\epsilon_1 + \epsilon_2$ equilibrium, meaning that truthtelling could be worse than $\epsilon_1 + \epsilon_2$ than truthful reporting but not more, so that agents who are indifferent to such small advantages would report truthfully. In this way, the mechanism can learn to tolerate heterogeneous agents with a reasonable amount of sample data.

5.2 PEER TRUTH SERUM FOR CROWDSOURCING (PTSC)

The peer truth serum presented in Chapter 3 has the distribution R of values as a parameter. When applied to a multi-task setting, we can learn the distribution R from a batch of submitted data. Furthermore, as we cannot reveal the distribution to the agents before they report their data, they cannot design collusive strategies around them. This is the idea underlying the *peer truth serum for crowdsourcing* (PTSC) [44], illustrated by Figure 5.2.

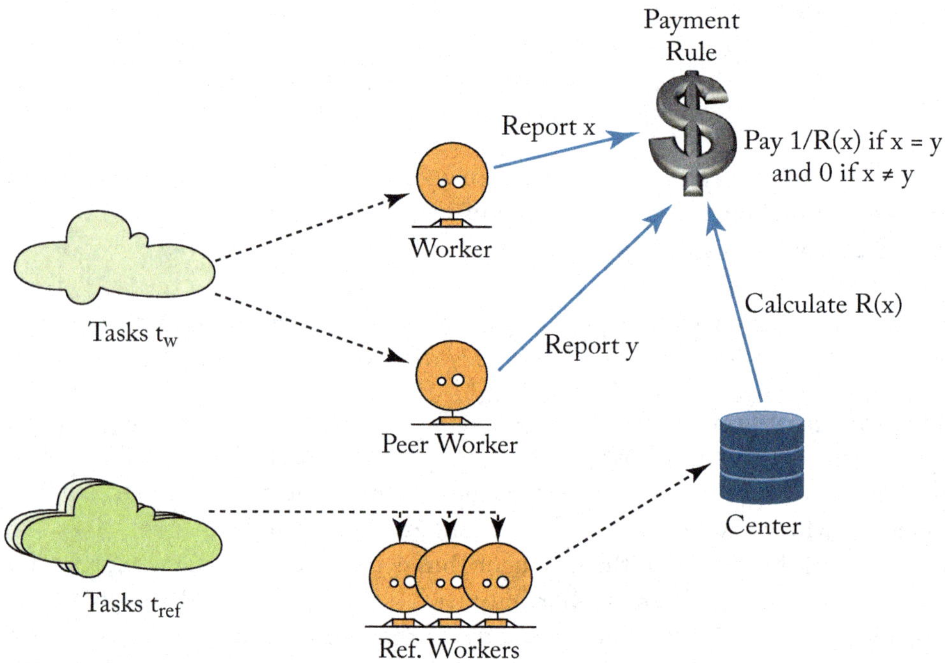

Figure 5.2: Scenario for the peer truth serum for crowdsourcing.

More specifically, in PTSC the distribution R is obtained as the histogram of reports from a set of many *similar* tasks, whereas the peer report is chosen from reports for the *same* task. The scheme is very intuitive and easy to understand: an agent should believe that his best guess at the prior distribution of peer reports is to let $P \simeq R$ (at least in the limit of infinitely many tasks), and that for its own task, $q(x)/R(x)$ is maximized for its own observation x_i. The result is Mechanism 5.2.

As an example to illustrate the PTSC mechanism, consider a set of tasks that have four possible answers a, b, c, and d. Assume that for a batch of ten tasks the center receives the answers shown in Table 5.1. This results in the overall distribution R of answers across all tasks:

Mechanism 5.2 The peer truth serum for crowdsourcing (PTSC) mechanism.

1. Center gives a set of similar tasks $\mathcal{T} = \{t_1, \ldots, t_m\}$ to agents $A = \{a_1, \ldots, a_k\}$ such that every task is solved by multiple agents; a_i reports data x_i.

2. For worker w, calculate the histogram of answers $r_w(x) = \frac{num(x)}{\sum_y num(y)}$, where reports by worker w are excluded.

3. For each task t_w carried out by worker w, select a peer worker j that has solved the same task. Reward agent a_i with a payment proportional to:

$$\tau(x_i, x_j) = \frac{\mathbf{1}_{x_i = x_j}}{r_w(x_i)} - 1.$$

When $r_w(x) = 0$, reward the agent with 0 (as there is no matching peer report).

Table 5.1: Answers received for the batch of tasks in the PTSC example

Task	Answers	g
t_1	b, a, a, c	a
t_2	b, b, b, a	b
t_3	a, a, b, a	a
t_4	a, d, a, a	a
t_5	c, c, a, b	c
t_6	d, a, d, d	d
t_7	a, a, c, a	a
t_8	b, b, a, b	b
t_9	a, a, a, a	a
t_{10}	b, b, a, b	b

Answer	a	b	c	d
Count	20	12	4	4
R	0.50	0.30	0.1	0.1

Within this batch, consider now an agent a_i who solves t_7 and has $x_i = a$. Suppose that it adopts the belief that its prior distribution is equal to R $(p(x) \leftarrow R(x))$ and that it updates its posterior correctly to reflect the distribution in the batch: $q(x) \leftarrow freq(x|a)$. This results in the following expected payoffs for different reporting strategies:

- honest, report a:
  $E[pay(a)] = \frac{0.75}{0.5} - 1 = \frac{1}{2}$,
- strategic, report c:
  $E[pay(a)] = \frac{0.1}{0.1} - 1 = 0$,
- random, report according to r:
  $E[pay([0.5, 0.3, 0.1, 0.1])] = 0.5 \cdot \frac{0.75}{0.5} + 0.3 \cdot \frac{0.1}{0.3} + 0.1 \cdot \frac{0.1}{0.1} + 0.1 \cdot \frac{0.05}{0.1} - 1 = 0.$

In fact, we can see that this is not an accident, but valid for all tasks. When we consider the probability of different answers across all tasks with the same answer, shown in Table 5.2, we see that for each task, reporting the correct answer has the highest probability of matching the peer, and the highest payoff!

Table 5.2: Probability of observing different answers, differentiated by the true answers of each task

Correct Answer		Observed Answer			
		a	b	c	d
a	$Count(a)$	15	2	2	1
	$freq(\cdot\|a)$	**0.75**	0.1	0.1	0.05
b	$Count(b)$	3	9	0	0
	$freq(\cdot\|b)$	0.25	**0.75**	0	0
c	$Count(c)$	1	1	2	0
	$freq(\cdot\|c)$	0.25	0.25	**0.5**	0
d	$Count(d)$	1	0	0	3
	$freq(\cdot\|d)$	0.25	0	0	**0.75**
	$Count$	20	12	4	4
	R	0.5	0.3	0.1	0.1

One issue with Mechanism 5.2 is that when some tasks have many more answers than others, the histogram can end up biased toward those answers. Thus, it may be desirable to collect the same number of samples from each task.

Given that the mechanism computes the exact distribution of peer answers, an equivalent incentive can be obtained as the expected value of the payment rather than the random match. Rather than paying:

$$\tau(x_i, x_j) = \frac{\mathbf{1}_{x_i = x_j}}{R_w(x_i)} - 1$$

with a randomly chosen peer report x_j, we compute the frequency $f_{tw}(x_i)$ and reward with

$$\tau(x_i) = \frac{f_{tw}(x_i)}{R_w(x_i)\sum_x f_{tw}(x)} - 1.$$

This eliminates the volatility caused by the random selection of a peer, and thus leads to more stable payments with the same incentives.

It is possible to show [44]:

Theorem 5.2 *Given a sufficient number of tasks, the peer truth serum for crowdsourcing (Mechanism 5.2) has cooperative strategy as a strict ex-post subjective Bayes-Nash equilibrium, and the payoff of this equilibrium is greater than that of all other equilibria.*

Furthermore, the peer truth serum for crowdsourcing has several useful properties.

- When truthful information requires costly effort, the rewards can always be scaled so that workers maximize their reward by investing maximal effort.
- Heuristic reporting, defined as giving answers according to a distribution that is independent of an observation, always reduces the expected payoff in comparison with truthful reporting.
- The larger the batch size, the more tasks are available to learn an accurate distribution R, and the weaker the self-prediction condition can be, i.e., the self-predictor (Equation 3.2) can be closer to 1.

However, the incentive-compatibility of PTSC hinges on the condition that agent's belief systems must satisfy the self-predicting condition, as defined in Definition 1.4. We now show a mechanism that does not require this condition.

5.3 LOGARITHMIC PEER TRUTH SERUM

The PTSC mechanism requires agents' beliefs to satisfy the self-predicting condition. However, the PTSC mechanism can be modified so that the self-predicting condition is no longer required, but instead truthfulness is achieved by rewarding the information content of reports according to a logarithmic information score. However, the price to pay is that the mechanism is guaranteed to be truthful only in the limit of infinitely many agents and reports.

More specifically, the *logarithmic peer truth serum* [45], shown in Mechanism 5.3, rewards a report x according to the logarithmic loss function $\log p(x)$. It measures the normalized frequency of the report x_i, $L(i)$ among the population of peers of the reporting agent, and rewards agents according to $\log L_i(x)$. To normalize the reward for random reporting to 0, it also measures the frequency in the overall population $G_i(x)$ and subtracts $\log G_i(x)$, so that the final

Mechanism 5.3 The logarithmic peer truth serum for crowdsourcing (LPTS) mechanism.

1. Center gives a set of similar tasks $\mathcal{T} = \{t_1, \ldots, t_m\}$ to agents $A = \{a_1, \ldots, a_k\}$ such that every task is solved by multiple agents; a_i reports data x_i.
2. For worker a_i and task T_m, calculate
   - the local histogram of peer answers $L_i(x) = \frac{num(x)}{\sum_y num(y)}$, obtained on the same task, and
   - the global histogram of answers $G_i(x) = \frac{num(x)}{\sum_y num(y)}$,

   where in both cases reports by agent a_i are excluded.
3. Reward agent a_i for answer x_i with a payment proportional to:

$$\tau(x_i) = \log \frac{L_i(x_i)}{G_i(x_i)}.$$

payment is:

$$\tau(x_i) = \log \frac{L_i(x_i)}{G_i(x_i)}.$$

Radanovic and Faltings [45] show that the expected payoff of the logarithmic peer truth serum can be expressed as the difference of two Kullback-Leibler divergences:

- the divergence between the distribution of the phenomenon given the observed value and the prior distribution without the report, minus
- the divergence between the distribution of the phenomenon given the reported and the distribution given the observed values.

The reporting strategy that maximizes the score makes the second divergence equal to zero, and this is only the case when the distributions are equal, which in turn requires reported and observed value to be identical.

With truthful reporting, the expected payment is equal to the positive term and a measure of the information gain by the measurement, so that it can be seen that the logarithmic PTS incentivizes meaningful measurements. It can be shown [45]:

Theorem 5.3 *Given a large enough number of peers, the logarithmic Peer Truth Serum (Mechanism 5.3) has truthful reporting as a strict ex-post subjective Bayes-Nash equilibrium; this equilibrum has a strictly higher payoff than all other equilibria except permutations of the truthful strategy.*

5.4 OTHER MECHANISMS

There are a few other mechanisms that have been proposed recently for a similar framework. Kong and Schoenebeck [4] also include a setting where agents provide multiple answers, and show a generalization of the correlated agreement class of mechanisms.

Kamble et al. [46] propose two mechanisms, for homogenous and heterogeneous agent beliefs. The mechanism for heterogenous beliefs is a variant of the PTSC mechanism with the same properties and requires the self-predicting condition to hold.

For homogeneous agent beliefs, they propose a truth serum that is truthful no matter how biased agent observations are, as long as they all share the same confusion matrix between observation and state of the phenomenon (see Section 1.3 and Figure 1.8 for an example). However, the mechanism requires that each group of very similar tasks, is evaluated by a large number of agents.

Mechanism 5.4 shows the mechanism. The clever idea is that the f_i reflect the degree that answers are self-correlated, and is obtained directly from the agents' answers. This allows the mechanism to learn the agents' reporting biases and use them in the incentives to ensure a truthful equilibrium. However, for this learning mechanism to be applicable, all agents must share the same reporting bias.

5.5 APPLICATIONS

5.5.1 PEER GRADING: COURSE QUIZZES

Peer grading is a technique for efficiently grading student exercises, in particular in large classes or online courses. Students grade the homeworks or exams of several classmates. Just as in crowdwork, there is no reason for students to carry out the task diligently; in fact there are even slight incentives to provide wrong results on purpose in order to influence the average score of the class.

In order to apply the PTSC mechanism, one needs to formulate the grading task so that it has a discrete and comparable answer space. The assignments to be graded were about programming and involved two types of questions: to fill in missing lines of code, or to find errors in a given piece of code. Each assignment could receive one of three grades: correct, and two different kinds of mistakes, and these were clearly explained to the participants (see Figures 5.3 and 5.4).

The quality of the grading was evaluated by comparing it against that of an expert grader, who was taken as the ground truth. Besides PTSC, the peer confirmation scheme of Huang and

Mechanism 5.4 The truth serum from Kamble et al. [46].

1. Center gives a set of similar tasks $\mathcal{T} = \{t_1, \ldots, t_m\}$ to agents $A = \{a_1, \ldots, a_k\}$ such that every task is solved by multiple agents; a_i reports data $y_{i,j}$ for task t_j.
2. For each each value $x_i \in X$:
   - for each task t_j, choose two reports $y_{l,j}$ and $y_{k,j}$ obtained for t_j by two different agents, and compute
   $$f_i^j = \mathbf{1}_{y_{l,j}=x_i} \cdot \mathbf{1}_{y_{k,j}=x_i},$$
   i.e., $f_i^j = 1$ if both $y_{l,j}$ and $y_{k,j}$ are equal to x_i, and 0 otherwise.
   - Compute
   $$\bar{f}_i = \sqrt{\frac{1}{m}\sum_{j=1}^{m} f_i^j}$$
   - Choose a scaling factor K and fix a payment
   $$r(x_i) = \begin{cases} \frac{K}{\bar{f}_i} & \bar{f}_i \notin \{0,1\} \\ 0 & \bar{f}_i \in \{0,1\} \end{cases}.$$
3. To reward agent a_i for answer $y_{i,j}$ on task t_j, pick another answer $y_{l,j}$ provided by a different agent for t_j. If the answers match, i.e., $y_{i,j} = y_{l,j}$, reward a_i with $r(y_{i,j})$, otherwise a_i gets no reward.

Fu [27] (constant reward if answer match) and a constant reward for every answer were used as rewarding mechanisms. That is, 48 students were divided into 3 equal groups of 16, where each group was rewarded by a separate mechanism; two students did not carry out the task. Each student graded four other quiz questions. Tables 5.3 and 5.4 show the error rates obtained by the different mechanisms, and the p-values obtained in a t-test comparing the answer distributions. We see that the PTSC mechanism has a significantly stronger effect than output agreement, almost halving the error rate.

5.5.2 COMMUNITY SENSING

Incentives per sensor Another evaluation was performed on the community sensing testbed for the city of Strassbourg that we already used earlier to evaluate the simple PTS mechanism, only now we discretize measurements to a scale of four values. In this application, using PTSC not only solves the problem that the mechanism does not know the correct R, but also eliminates

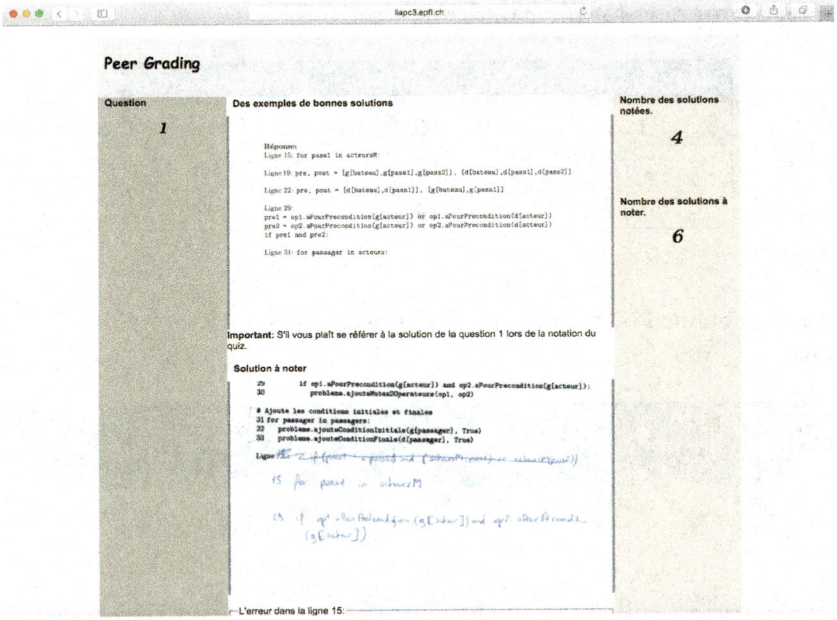

Figure 5.3: Peer grading experiment: assignment of the first type.

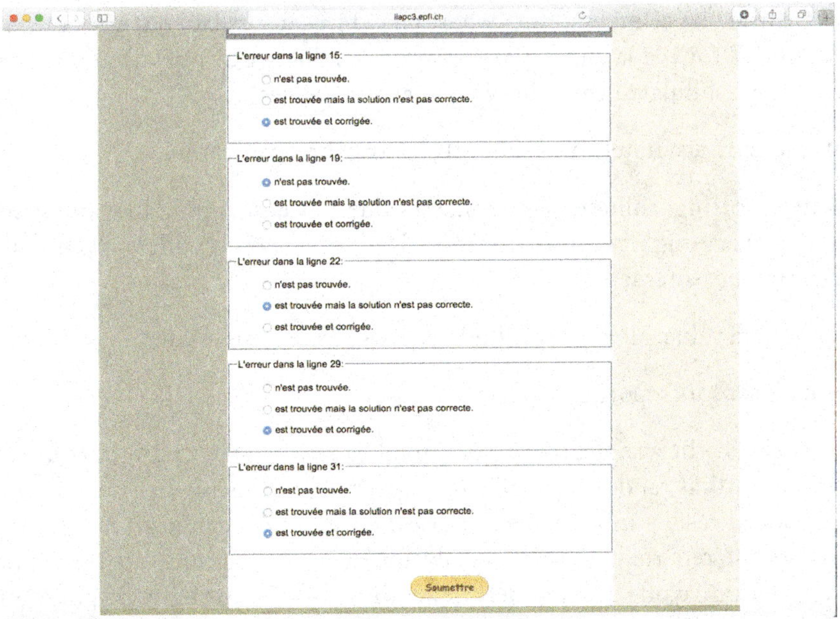

Figure 5.4: Peer grading experiment: assignment of the second type.

Table 5.3: Average error rates for three different mechanisms

Mechanism	Num. Students	Error Rate (%)
PTSC	16	6.88
OA	16	10.48
Constant	14	11.98

Table 5.4: p-values obtained by a t-test for the answer distributions obtained by the three different mechanisms

Mechanism	PTSC	Peer Consistency	Constant
PTSC	–	0.0255	0.0497
OA	0.0255	–	0.5566
Constant	0.0497	0.5566	–

the possibility for agents to obtain gain by colluding to always report the value with the highest payoff.

As in the earlier experiments, the scenario has 113 sensors laid out over the city of Strassbourg (see Figure 3.8 for the layout). The continuous value space was discretized into four different values, and we simulated the following agent strategies.

1. Honest: measure accurately and honestly report the observation.
2. Inaccurate reporting: collude on a reduced value space, mapping low and medium to low, and high and very-high to high. This models agents who spend less effort and thus obtain an inaccurate measurement.
3. Collude on one value: all agents collude to report the same value.
4. Random: all sensors report randomly.

Figure 5.5 shows the average reward obtained by sensors in each of the four strategies. We can see confirmation that random or collusive reporting, not using any information about the measurement, indeed results in an average reward of zero. We can also see that accuracy pays off, as the strategy the reports with lower resolution also has a significantly lower reward.

However, sensor payoffs differ widely depending on how strongly the pollution levels vary at the sensor's location. Figure 5.6 shows the average rewards obtained by each sensor side by side. The figure shows two strategies: the honest strategy in blue, and the random strategy in red.

We can see that for each and every sensor, the average payoff of the honest reporting strategy is significantly higher than that of the random strategy, so that the scheme works not only in expectation, but in every individual case.

Note also that there are big differences in expected payoff depending on sensor location: some sensors obtain much higher rewards than others. This is due to the fact that the PTS mechanism gives higher rewards for measuring at uncertain locations. It is very useful to incentivize self-selection so that agents position their sensors where they provide the most new information to the center.

For comparison, Figure 5.7 shows the distribution of rewards for both strategies for sensors that randomly change locations among the 116 positions; they could be considered mobile sensors that move around the city. We can see that rewards are more evenly distributed, while maintaining a very consistent and high advantage for truthful sensors.

As mentioned in the earlier sections, one can relax the requirement for the self-predicting condition to hold, by using Log-PTS. However, this relaxation comes at a certain cost—the

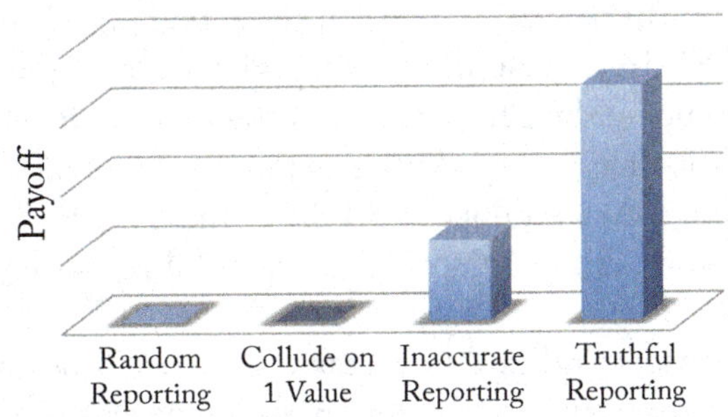

Figure 5.5: Average payoffs observed in the Strassbourg simulation for different agent strategies.

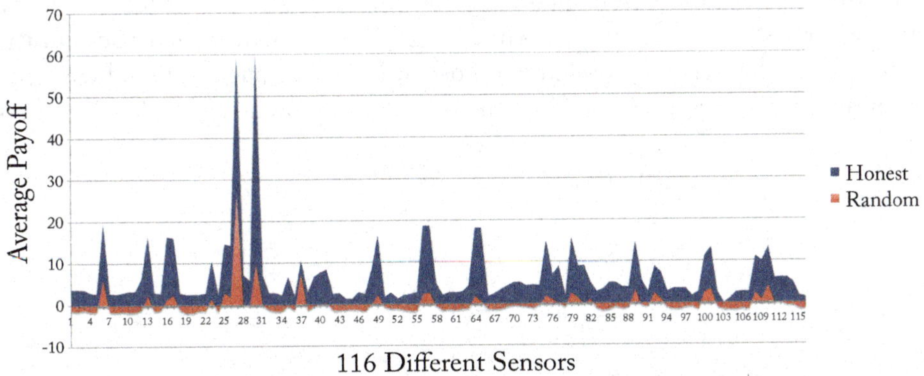

Figure 5.6: Average reward for each of the 116 sensors when they are static.

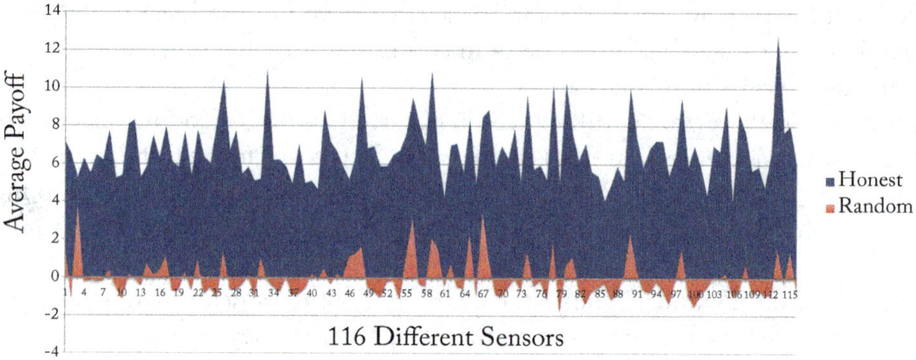

Figure 5.7: Average reward for each of the 116 sensors when they are mobile.

new requirement is that the number of peers is large, which in the community sensing scenario implies that each sensor has a larger number of neighbors—this is in contrast to PTSC, which requires only one peer. In particular, one can show that Log-PTS indeed exhibits similar incentive properties as PTSC in community sensing, given a large population of peers [45].

It is interesting to compare the performance of the two mechanisms when the population of sensors decreases, both in the term the total population size and the number of available peers.[1] Figure 5.8 shows the scores produced by PTSC and Log-PTS for different numbers of sensors and peers and for three different strategies: truthful reporting, colluding on one value and inaccurate reporting.

As the number of sensors and peers decreases, Log-PTS becomes less robust to uninformed and inaccurate reporting, that is, the difference between the average scores for truthful reporting and the misreporting strategies decreases. Once the total number of sensors is around 40 and the number of peers is around 7, Log-PTS no longer manifests its theoretical properties. Instead, the highest paying strategy is collusion. In contrast, PTSC preserves its properties under much greater stability. Therefore, while Log-PTS is applicable to community sensing scenarios that have a dense network of crowd sensors, PTSC is much more robust to deviations from this condition, and applies even when the network of crowd-sensors is relatively sparse.

[1]Both of the mechanisms use all the available peers to calculate the score. In the case of Log-PTS, the reports of peers are used to calculate quantity L_i. In the case of PTSC, we simply average scores across all the peers.

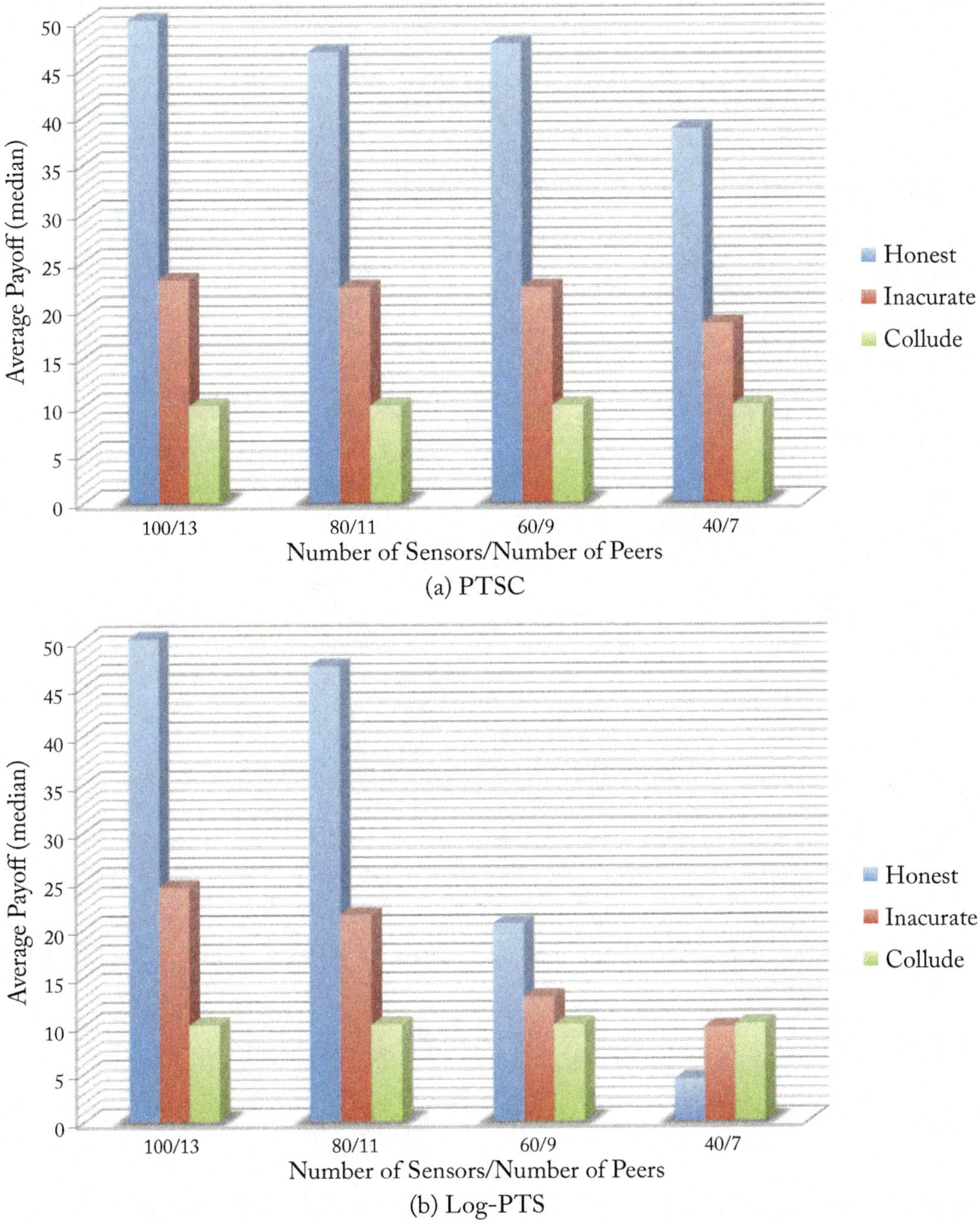

Figure 5.8: The robustness of PTSC and Log-PTS in terms of the sensors' population size.

CHAPTER 6

Prediction Markets: Combining Elicitation and Aggregation

When multiple agents provide predictions or estimates of a value, we would like to weigh this information according to the confidence that they attach to it. However, this confidence must also be elicited in some way—agents may want to have a strong influence and thus overclaim their influence. The idea of a *prediction market* is to make agents take *risks* in proportion to their confidence, so that having a larger influence on the aggregate information requires taking a larger risk.

A prediction market is modeled on a financial market, where agents express their predictions by buying *securities* that will pay off once the ground truth g becomes known. When eliciting predictions for a phenomenon X with values $x_1, \ldots, x_N$, there will be one security for each outcome x_i that pays off \$1 if the ground truth turns out to be equal to x_i, and nothing otherwise.

At every point in time, each security has a market price $\pi(x_i)$. When agents buy security x_i, the price will rise, and thus the consensus probability will increase. Whenever the agent believes that $\pi(x_i) < q(x_i)$, it is rational for the agent to buy the security, since the price is lower than the expected payoff $q(x_i)$. Likewise, when the price is higher than $q(x_i)$, a rational agent would sell the security. Therefore, the market price is in competitive equilibrium when $\pi(x_i)$ is a consensus probability estimate for $\Pr(g = x_i)$.

The principle here is that a bigger investment buys a bigger influence, but also increased risk if the result is not as expected. When agents have a limited budget, or are risk-averse, they will invest their resources where they are the most confident. Thus, a prediction market tends to give higher weight to more confident agents (Figure 6.1).

An example of a prediction market that has been in operation for many years are the Iowa Electronic Markets [47]. Figure 6.2 shows an example of the price evolution of the market for the two candidates in the 2008 U.S. presidential election. In all recent elections, this market was more accurate than opinion polls.

It is easy to see that there are many other scenarios where such a market could be applied: predicting when a project will be completed, whether a new product will be a success, or whether there will be more demand for oil are examples. However, a major problem is that trading secu-

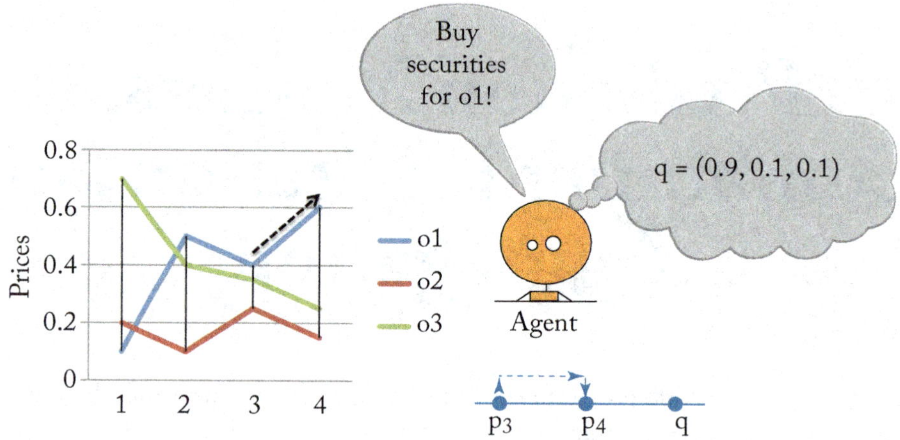

Figure 6.1: Prediction markets, seen from an agent's perspective.

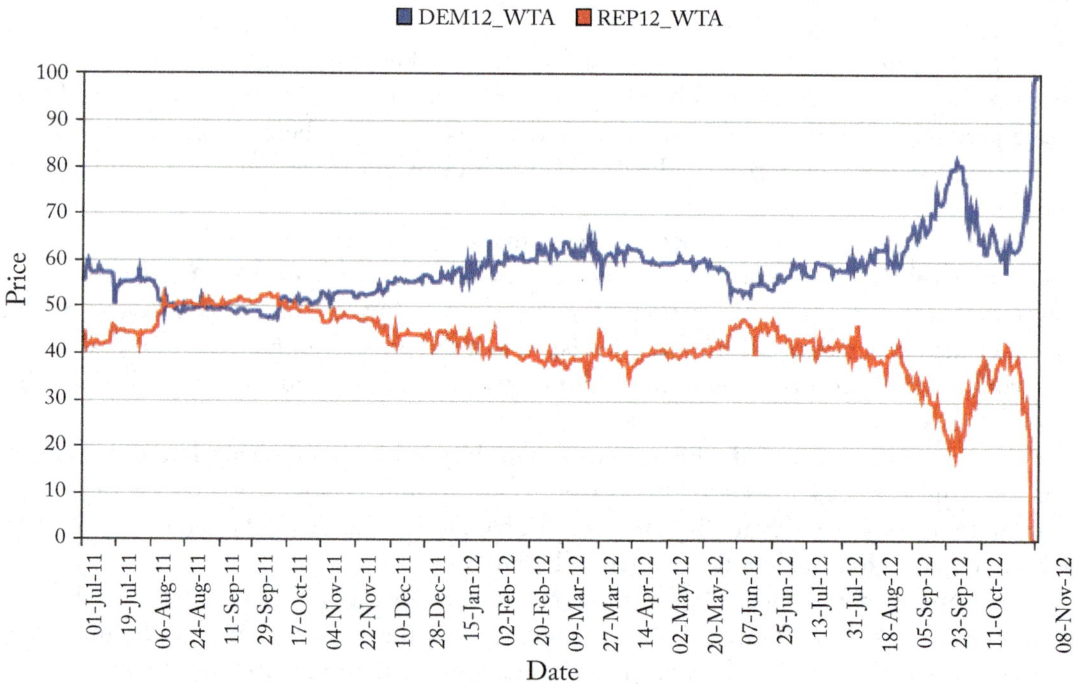

Figure 6.2: Price evoluation for the democratic (blue) and republican (red) candidate in the 2008 U.S. presidential election.

rities requires many active participants. For a presidential election, there is always much interest and so anyone willing to buy or sell a security will find someone to trade with in a short time. However, for more specific questions there may not be many other traders, and thus the market may be tedious.

For this reason, most prediction markets use artificial counterparties called *automated market makers*. These are agents that are committed to trade at any time and with any counterparty at some price. In practice, they simply generate or eliminate the securities that they trade. The main question is what should be the price for selling and buying them?

A clever answer to this question has been found through the use of scoring rules as we saw them in the previous chapter [52, 53]. The idea is to score the quality of the distribution expressed by the current market prices by a proper scoring rule, applied against the true outcome when it will become known. The reward for an agent who moves the price by buying or selling securities should then be equal to the amount by which she changes this score: if it is improved, the agent should expect a positive payoff, otherwise a negative one. In this way, the score of the distribution is distributed in a fair way among the agents that participated in defining that distribution.

Such a principle can be implemented by choosing the price function based on the scoring rule. Consider first a simple market with one type of securities that pay 1 if x_i occurs, and 0 otherwise. Agents can trade any quantity of this type of security. Let $\pi(n)$ be the price for buying/selling an infinitesimally amount, given that n securities are held by other agents in the market.

We design a market maker based on the logarithmic scoring rule, which has become very common for prediction markets. Assume that an agent believes that true probability of outcome x_i is $\pi^*(x_i) > \pi(x_i)$. Therefore, it buys m securities and makes the price increase to some $\pi(n + m) = \pi' > \pi(n)$. What should this price be?

Considering the score-sharing principle outlined above, its profit should be determined by the scoring rule $Sr(Pr, g)$ to be $Sr(\pi', 1) - Sr(\pi, 1)$ if the outcome is indeed x_i:

$$m - \int_n^{n+m} \pi(\mu) d\mu = Sr(\pi(n+m), 1) - Sr(\pi(n), 1).$$

Taking the derivative with respect to m, we obtain:

$$(1 - \pi(n)) = \frac{dSr(\pi(n))}{dn} = \frac{dSr}{d\pi}\frac{d\pi}{dn}.$$

Using the logarithmic scoring rule, $Sr(\pi) = b \ln \pi$, we obtain the price function

$$\pi(n) = \frac{e^{n/b}}{e^{n/b} + 1},$$

where b is a *liquidity parameter* that determines how much each share moves the price. If there are many participants, or they are willing to invest substantial amount of money, then the liquidity parameter should be high. On the other hand, if it is too high, then agents cannot move the prices enough to obtain a reasonable estimate.

Besides the difficulty of setting the right liquidity parameter, an issue with automated market makers based on the logarithmic scoring rule is that the price of a security can never reach 1 and reflect a certain outcome: as the price approaches 1, making gains requires buying huge numbers of securities, and thus taking huge risks! Therefore, such markets are most suitable for problems with quite uncertain outcomes.

If we consider that agents arrive one by one and each buy or sell shares until the price matches their own opinion, the market will fluctuate and never actually aggregate information from multiple participants. However, if traders consider that they have only observed a limited sample of the phenomenon, they should take the current market price into account when forming their own opinion.

If they use a Bayesian update as in Equation (1.1), they will form a weighted average of their own opinion and that of the market. Studies such as that of Abernethy et al. [54] show that when agents believe the observed samples to be distributed according to an exponential distribution, and perform frequentist belief updates, the market will indeed have equilibria where agents agree on a common average while each having a slightly different opinion based on their own sample. Under these conditions, the prediction market thus obtains the same outcome as a center that aggregates honest reports of observations, such as in peer consistency.

Automated market makers can be constructed with any proper scoring rule using the same principle as shown above. The reason why the logarithmic scoring rule is often used is that it is the unique proper scoring rule that admits consistent prices for *combinatorial* predictions. For example, we could imagine a market that would not only have securities for the presidential election, but also for the congressional elections, and for combinations of both election outcomes. For example, there would be a security for the event "the democratic candidate wins the presidential election but the majority in congress goes to the republican party." If an agent buys a security for just the presidential election, this should also influence the price of the combined securities. The logarithmic scoring rule correctly models this; for more details see Hanson [52].

Prediction platforms We already mentioned the Swissnoise platform in Chapter 3. It was operated with artificial money as a prediction market with a logarithmic scoring rule market maker. The artificial money accumulated by each participant was shown in a leaderboard, and each week the participant with the biggest profit was awarded a small gift certificate.

Questions were limited by the fact that prediction markets require a verifiable outcome, and so hypothetical questions ("What would happen if…") could not be asked.

Figure 6.3 shows a screenshot of the interface, with the example question of what team would win the 2014 FIFA soccer world cup. Only four teams were remaining at the time of this screen shot, and the graph shows the price evolution in recent days.

To place a bet, a participant would choose one of the available securities. Figure 6.4 shows this for the example question, where the participant already has shares in the outcome "Ger-

Figure 6.3: Example question on the Swissnoise prediction market.

many" (who was the eventual winner of the tournament). Once a security is chosen, the participant interacts with the market maker who determines the price. This is the part that is the most abstract, and thus the hardest to understand. To support it, Swissnoise showed an interactive slider where one could see the price evoluation associated with buying a certain number of securities (see Figure 6.5). By moving the slider, it is possible to see the cost of buying a certain number of shares of the security, according to the scoring rule used by the market maker.

The design of the Swissnoise platform was deemed quite attractive by users, and resulted in a steady participation. However, we noted two major difficulties with such prediction markets. The first is that it is very difficult to fix the right value of the liquidity parameter. If it is set too small, there are huge price swings and the market does not estimate any meaningful probabilities, and this happens easily for questions gain in popularity and attract more participants that trade. Garcin and Faltings [55] analyze the optimal liquidity parameter for three different questions that ran on the platform, and shows that they were vastly different: 25, 480, and 1,250 for the three different questions. As the parameter has to be kept constant while running the market, it is very difficult to ensure that it is set to a good value.

Figure 6.4: Placing a bet on the Swissnoise platform.

The second issue is that once the market reaches a very high probability for a particular outcome, there is little interest for participants to continue to hold the corresponding security: selling it at the current price results in almost the same profit as what could be gained by holding on until the end. On the other hand, the profit that is gained from selling securities early can be invested on other questions where answers are not as clear yet, and result in much higher profit. To the market, however, such selling cannot be distinguished from the participant having changed her mind about the outcome, and this further contributes to price instability.

As a result, Garcin and Faltings [55] observed that prediction markets tended to develop wild price swings, making the theoretical idea of predicting the probability of different outcomes almost meaningless. An example is shown in Figure 6.6, which shows the evolution for the prediction of the referendum on Scottish independence in 2014.

Abernethy and Frongillo [50] show how to use the mechanism of a prediction market for incentive-compatible collaborative learning, where agents make predictions on different hypotheses for the learning outcome and are scored according to how well they predict perfor-

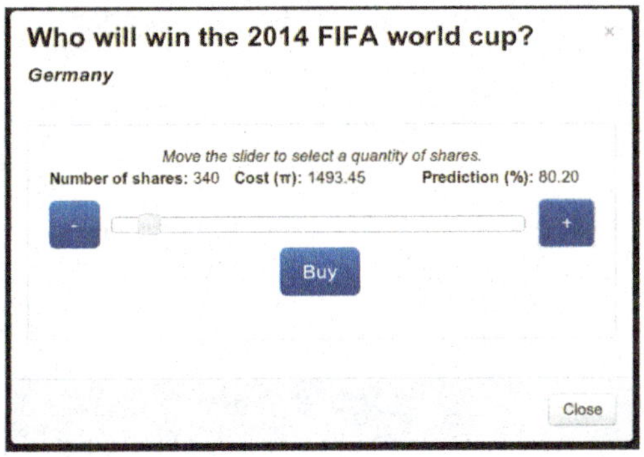

Figure 6.5: Interaction with the market maker.

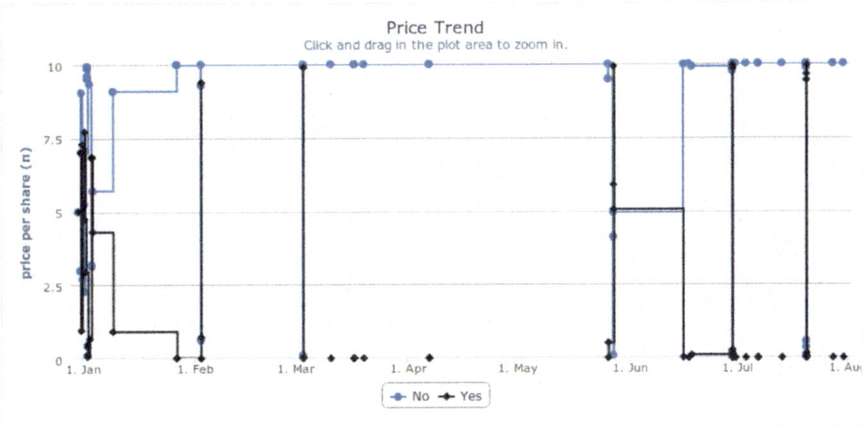

Figure 6.6: Price evolution for the question of whether the referendum on Scottish independence will succeed, using a market maker with a logarithmic scoring rule.

mance on test data. This mechanism incentivizes agents to honestly collaborate to form the best consensus learning outcome.

Frongillo et al. [51] show how to infer the confidence that an agent has about a prediction from his answer itself. This confidence can then be used to aggregate the answers into a compromise aggregate that reflects the confidence of each agent.

As a concluding remark, notice that this section provided only a brief overview on prediction markets. In general, there is the vast literature on prediction markets, covering different aspects of information elicitation using the prediction market framework. This section, nevertheless, provides a basic insight in this framework from, relating it to the problem of aggregating elicited data.

CHAPTER 7

Agents Motivated by Influence

An important problem in using contributed data is that some agents may be insensitive to monetary incentives and not adopt the cooperative strategies we expect. This mainly happens in two cases. The first is *faulty* agents, who provide incorrect data in spite of their best efforts to be cooperative. Another case is *malicious* agents, who want to insert fake data to influence the aggregate for ulterior motives, for example to hide pollution or to influence a decision that is taken based on the data. Following the practice widely adopted in fault-tolerant computing, we consider faulty agents as malicious as well in order to obtain worst-case guarantees.

For malicious agents, the game is different: in addition to the *cost* for obtaining the data, agents also care about their *influence* on the learning outcome obtained by the center. As this influence is usually much more important than the cost, and to simplify the techniques, we assume that they *only* care about the influence on the outcome.

To analyze this influence, the aggregator—more generally, the learning algorithm—used by the center is crucial. By default, we will consider Bayesian aggregation, which in the case of a simple data point corresponds to forming an average.

Like in the case of monetary incentives, we distinguish cases where the ground truth becomes known, and cases where the ground truth is never known.

We first consider the case where the ground truth can be verified or becomes known at a later time. This comparison with the ground truth can be used to maintain a *reputation* that determines the influence that data reported by the agent has on the aggregate result. In this way, it is possible to bound the influence that an agent can have on the learning outcome, through a process called the *influence limiter*.

The case where the ground truth is never known is of course more challenging. In general, this means that data agregation becomes a kind of negotiation where different data providers attempt to influence the learning outcome to their most preferred result, and the actual truth actually does not even matter. This case can be analyzed using techniques for the problem of *social choice*, but since it no longer relates to data we do not consider it further here.

For some cases, there are incentive-compatible mechanisms where an agent gets the best possible influence by reporting the data that would fit his own preferred model. This has interesting applications when agents may each have a different view on the data, and would be best served if the center adopted a model that is as close as possible to their own. For example, if agents report on the quality of the food served in a restaurant, they could consider that items they do not like will disappear from the menu while those they do like may appear more often. We will consider this case in a further section.

7.1 INFLUENCE LIMITER: USE OF GROUND TRUTH

While we cannot eliminate the occurence of such agents, we will now show a way that we can *limit* their negative *influence* on the learned model through a reputation system. It is based on an idea originally developed to combat fraud in recommender systems [56], and we keep the name *Influence Limiter* that was introduced there.

Figure 7.1 shows the scenario assumed in the influence limiter. Agents report data sequentially over time, in the order of Agent 1, then 2, then 3. The center uses this data to produce new aggregate models 1, then 2, then 3, that are based on all data received up to that time. The models could be averages, or complex models obtained through machine learning algorithms. They can be used to estimate the value of all variables associated with the phenomenon, possibly by interpolation.

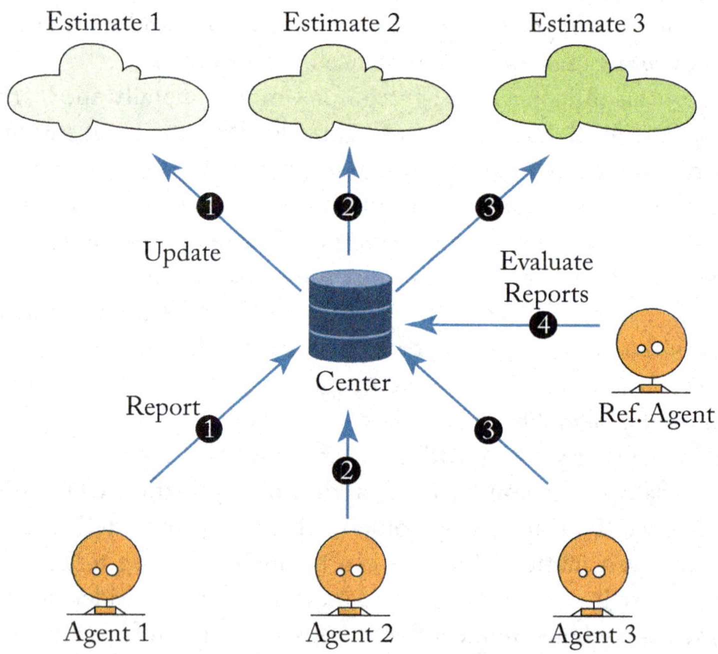

Figure 7.1: Data aggregation setting assumed for the Influence Limiter.

Occasionally, the center obtains the ground truth $g_t(X)$ for some observation X of the phenomenon, for example because a prediction can be verified, or it obtains a trusted measurement. This allows to evaluate the quality of its aggregate models by comparing the estimate it provides for this variable. It uses this opportunity to evaluate how much the data received from agent i actually improved the quality of the aggregate model, and thus to evaluate the quality of the data itself.

More precisely, suppose that the model before incorporating the data received from agent i predicted a distribution $p(X)$ for the observation X, and this changes to $q(X)$ after incorporating its data. The center evaluates the improvement in the model using the proper scoring rule Sr on the reference measurement g_t of X:

$$score_t = Sr(q, g_t) - Sr(p, g_t) \in [-1, +1].$$

Thus, we obtain a similar mechanism as in prediction markets, where each agent obtains a reward proportional to its contribution to the quality of the learned model through the data it provided, and we could in fact use this as an incentive scheme as well.

However, in the influence limiter, we are interested in using this information to limit the influence that an agent can have on the learning outcome. We do this by assigning each agent a *reputation*, and making its influence on the model depend on this reputation.

We define the influence of an agent as follows.

Definition 7.1 Consider an agent i at time t and let us denote the aggregation output prior to the agent's report by $o_{-i,t}$ and the aggregation output posterior to the agent's report by $o_{i,t}$. The *influence* of agent i is defined as $influence_{i,t} = Sr(o_{i,t}, g_t) - Sr(o_{-i,t}, g_t)$. The *total influence* of an agent is the sum of its influences over different time periods.

Notice that this definition also allows an aggregation mechanism to discard an agent's report, in which case the influence of the agent is equal to 0. The aggregation mechanism may decide whether or not to include an agent's report stochastically, in which case the influence is the expected value of the quality change.

Our primary goal is to design an aggregation mechanism that limits the negative influence of an agent, which by Definition 7.1 means that the total influence of an agent should be lower bounded. This goal is fairly easy to achieve if we simply discard all of the reports, however, in doing so, we also loose the information coming from agents that positively influence the learning procedure. Therefore, we also ensure that the information discarded from agents who have a positive influence is bounded from above. To quantify the performance of an aggregation procedure in this context, we introduce the measure of information loss.

Definition 7.2 Consider an agent i who is expected to have a positive influence. The *information loss* of an aggregation mechanism for potentially discarding agent i's reports is defined as $\mathbf{E}[score_{i,t} - influence_{i,t}]$. The *total information loss* associated with agent i is the sum of the information losses over different time periods.

Reputation systems Reputation is a well-known factor for preventing misbehavior in human society. The fact that uncooperative behavior leads to bad reputation and thus will be punished in the future counteracts the natural temptation for fraud and malicious behavior that is omnipresent in many interactions. This *reputation effect* is also widely exploited in distributed computation to reward cooperative agents.

A general way of using reputation in data aggregation is *thresholding*: submitting data requires a reputation that exceeds a minimal threshold, and otherwise will just be ignored. The most common technique used today, the β reputation system [57], computes an agent's reputation at time t as the fraction of the "good" interactions α_t up to time t vs. the "bad" interactions β_t:

$$rep_t = \frac{\alpha_t}{\alpha_t + \beta_t},$$

where $\alpha_t = \alpha_0 + \sum_{s \in \{scores_\tau > 0\}} |s|$ and $\beta_t = \beta_0 + \sum_{s \in \{scores_\tau < 0\}} |s|$.
However, a malicious agent can easily manipulate this scheme and exert arbitrarily strong influence on the aggregate model [58], by iterating the following steps:

- provide good data that does not change the model, and thus build up a good reputation and
- use this reputation to insert bad data that *does* change the model.

Thus, while such schemes are effective against faulty agents, they do not solve the problem of malicious agents.

Stochastic influence limiter The attack scheme we just described shows that reputation should not be based on the *quality* of the data provided, but on the *influence* that it has on the learned model. This is precisely what the scoring scheme we outlined earlier provides.

In the influence limiter, we compute an agent's reputation through an incremental reputation update based on the score it obtains for the data it contributed:

$$rep_{t+1} = rep_t \cdot \left(1 + \frac{1}{2} \cdot score_t\right).$$

Using this update function, the score has a stronger effect than in the β-system that is commonly used: reputation increases or decreases exponentially fast. Agents start out with a common initial reputation rep_0 which is updated every time they submit data. Note that the update could happen somewhat later if no observation that allows evaluation of the model quality is available at that moment.

We also replace the thresholding scheme by a stochastic information fusion: rather than accepting all data when reputation is above a threshold, we accept data from an agent with reputation rep_t stochastically with probability $\frac{rep_t}{rep_t+1}$.

Mechanism 7.1 The stochastic influence limiter mechanism.

Initialization: all agents a_i have an initial *reputation* $rep_0(i) = \rho$. The center has an initial model M_0 of the phenomenon.

At each time period t, the center receives reports from agents in sequential way and updates the model as follows:

1. Center initialize model $M_t = M_{t-1}$.
2. Center receives report x_i from agent a_i. It constructs a tentative updated model M_t^i by incorporating x_i.
3. Center sets $M_t^{-i} = M_t$, and updates $M_t = M_t^i$ with probability $\frac{rep_t(i)}{1+rep_t(i)}$.
4. After a reliable data x_g point is received, center evaluates model M_t^i and obtains a score $S(M_t^i, x_g)$; it computes the score of agent a_i's report as $score_i = S(M_t^i, x_g) - S(M_t^{-i}, x_g)$.
5. Center updates the reputation $rep(i)$ as $rep_{t+1}(i) = rep_t(i)(1 + \frac{1}{2}score_i)$.

For the resulting Mechanism 7.1, we can show two important properties [59].

Theorem 7.3 *Using the stochastic influence limiter mechanism, the total expected influence that any agent can have on the aggregate model is lower bounded by* $-2 \cdot rep_0$ *(the total influence of an agent is not overly negative). Furthermore, the resulting information loss is upper bounded by a constant (see [59] for the exact expression).*

Thus, the influence of any malicious agent is limited. However, this limitation comes at a price, since each data item is discarded with a certain probability. Therefore, the second property is also important.

Reducing the need for reference answers The stochastic influence limiter requires that we can assess the myopic influence of every data element on the quality of the model. This requires access to correct reference answers to score the quality of the model. Are there clever techniques for reducing the amount of reference answers that are used?

In the presented model, any data point that is stochastically relevant to the reported data can be used as a reference answer, and thus we can reduce the number of different data points as long as we maintain sufficient coverage of the data that is provided. However, with few reference answers the stochastic relevance will become weaker and so the evaluations will become more volatile.

Another possibility is to apply the reputation scheme only to the subset of the data for which a relevant reference answer is available, similar to gold tasks in crowdsourcing. Since agents do not know which data is being scored, they cannot devise a strategy to counter the scheme, and so we can expect similar guarantees to hold provided we can keep the choice of evaluations secret.

For crowdsourcing platforms, Shah and Zhou [60] propose a scheme where every correctly answered gold task doubles the bonus, while even just one wrong answer makes it drop to zero. This has the added effect to motivate workers to skip a task when they are not sufficiently confident about the answer. The scheme is very similar to the influence limiter, except that the evaluation is not based on the influence on the model and it is applied to a finite batch of tasks. It is shown to be unique in giving the lowest possible payoff to agents that provide only wrong data. However, it does not give any guarantees on the influence on the model.

In Shah and Zhou [61], the same authors extend this scheme in the following way: workers are confronted with the answer of a peer worker on the same task, and allowed to change their answer accordingly. They show in a simulation that this points workers to inadvertent mistakes and thus increases the quality of the result.

Steinhardt et al. [62] derive a scheme that minimizes the number of reference answers required to isolate the best quality answers in a crowdsourcing setting. This is an alternative approach to the influence limiter. It provides relative guarantees of identifying the best answers to some percentile, but no absolute guarantees on the accuracy of either the answers or the resulting model.

Application in community sensing To evaluate the practical performance of the influence limiter, Radanovic and Faltings [59] performed a simulation study on the scenario of community sensing, shown in Figure 7.2.

They compared two reputation systems:

- CSIL—stochastic influence limiter and
- BETA—beta reputation system.

The performance of the community sensing influence limiter was evaluated on the pollution model of the city of Strassbourg that we already explained in the Section 3 (Figure 3.8). The focus was put on how to counter malicious agents using the community sensing stochastic influence limiter (CSIL), in a (simulated) scenario with 40 mobile crowd-sensors, out of which 75% of them malicious with the following strategies.

1. Vary—initially report truthfully, and then start reporting only low level of pollution.
2. Deceive—report truthfully when the reputation is below a certain threshold, and otherwise report low level of pollution.
3. Vary and deceive—initially report truthfully, and then start reporting according to *deceive*.

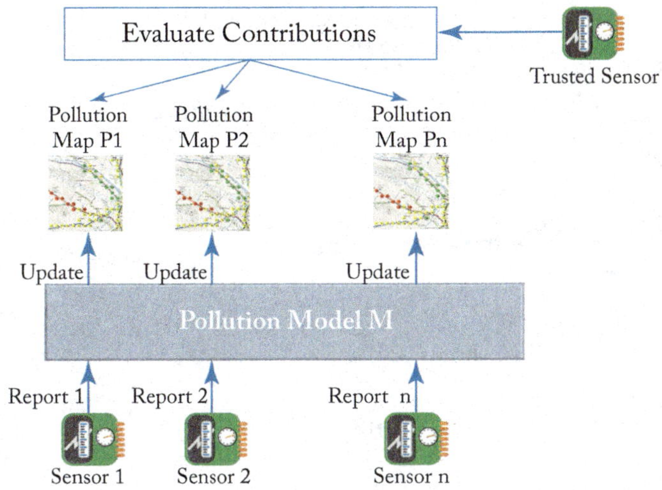

Figure 7.2: Community sensing scenario.

4. Cover—a more sophisticated version of the *vary and deceive* strategy where malicious sensors only misreport when they measure high enough level of pollution.

For the influence limiter, we used the quadratic scoring rule and compared against the β-reputation system that is commonly used. Figure 7.3 shows the empirical performance of the reputation system against these strategies. We see that the β-system is effective to prevent the *vary* attack strategy, but not against any of the others. In fact, its empirical performance is often worse than the theoretical worst-case bound obtained for the influence limiter. On the other hand, the empirical performance of the influence limiter is often well below the theoretical bound.

7.2 STRATEGYPROOF MECHANISMS WHEN THE GROUND TRUTH IS NOT ACCESSIBLE

In this section, we consider the case where the ground truth is not accessible at all. When all agents are interested in the learning outcome obtained by the center, there is no reason why the truth should even matter; agents' reports are opinions rather than data. However, we may consider truthfulness in the sense that agents truthfully report their most preferred data. For example, if the center is asking agents how useful the different outcomes of a decision would be to them, the center may want their answers to truthfully represent these preferences.

Deckel et al. [64], Meir et al. [65], and Meir et al. [66] consider this question for the case of regression and classification, respectively, and show that for certain cases, incentive-compatible mechanisms are possible. Notice that these two approaches differ from the approach

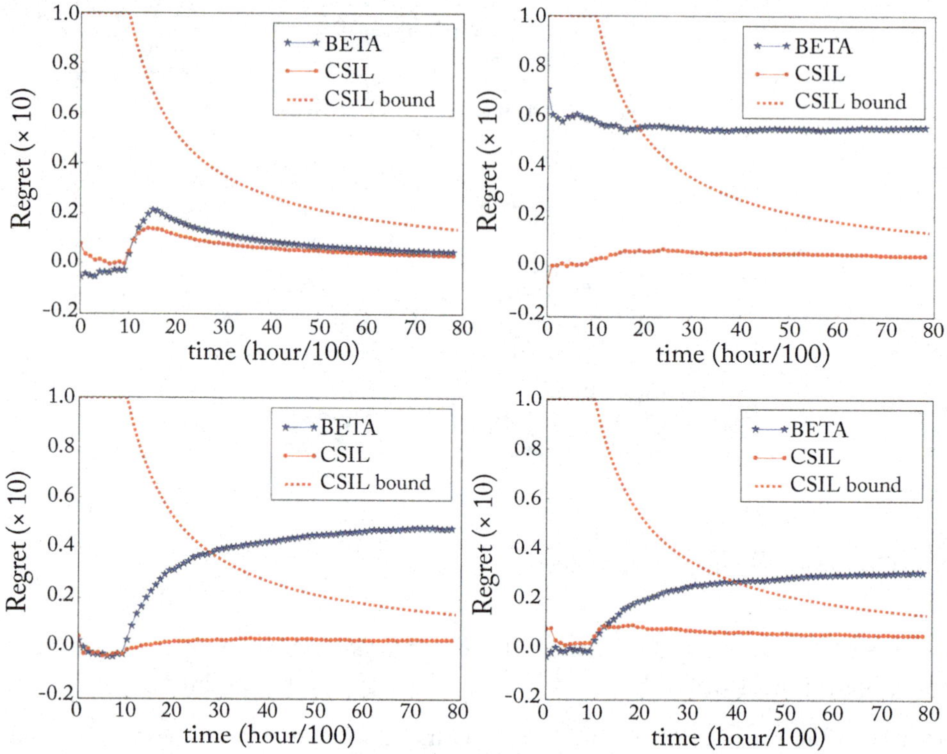

Figure 7.3: Empirical performance of the community sensing influence limiter compared to the theoretical worst-case bound and the empirical performance of the β-reputation system.

we described in the objective: the focus is put on optimizing the agents' preferences, which differs from the goal of eliciting correct data.

For regression, Deckel et al. [64] analyze a scenario where each agent has the desire to minimize an individual loss function of the model prediction and its own preferred value. The center is assumed to minimize the average loss of all agents that report data.

For the case where each agent is only interested in a single data point, it turns out that the setting is incentive-compatible only for the absolute loss function, i.e., where each agent would like to minimize the absolute difference between its most preferred value and the model prediction. Note that the model that minimizes the average linear loss is to let the prediction be the median of the values reported by the agents.

Another variant of this model is to assume that agents are interested in minimizing the expected loss over a range of points, assuming a uniform distribution. Consider that the loss function is again linear, and that the model to be learned is a constant function that returns the same value for every point. Note that with this restriction, there often is no model that fits an agents' preferences exactly. It turns out that agents have an interest to project their most preferred values onto the allowed model, i.e., report the same value for all data points, and report those values. To eliminate this incentive, the center can apply this projection for them, and then calculate the best combined model for all of them. It turns out that this mechanism is incentive-compatible, and efficient with an approximation ratio of 3, i.e., the solution is never worse than a factor of 3 from the true optimum.

For classification, Meir et al. [65, 66] analyze a scenario where each agent labels a set of points with positive and negative labels, and this information is used to learn a classifier that assigns the same class to all of the points—a similar restriction to the constant function assumed just above. Note that in this case, each agent prefers the positive class if it labels more points as positive than negative, and the negative class otherwise. The mechanism they propose labels the agents accordingly as positive and negative agents, and obtains the common label by majority vote where positive agents vote for a positive classification for each of their examples, and likewise for the negative agents.

Meir et al. [65, 66] show that this mechanism is incentive-compatible, i.e., agents will report their labels truthfully, and that the resulting classifier is a 3-approximation to the best classifier. The same paper also shows that a randomized algorithm can achieve a 2-approximation for the same setting.

Finally, we briefly mention the literature that goes in the direction of analyzing information aggregation without direct verification. In general, it is inconceivable to have a robust mechanism that allows arbitrary manipulation strategies of malicious users and outputs accurate aggregates. In fact, in the most general case, one can only hope for outputting a list of aggregates containing an accurate aggregate. This paradigm is discussed in more details by Charikar et al. [68], who provide a clustering based method whose goal is to detect a group of reports that come from reliable sources. Charikar et al. [68] prove the soundness of this approach and further relate it to a setting in which a mechanism designer can use a few reliable reports to differentiate which of the possible aggregates is correct. One can show that only a bounded number of reliable ratings are needed in the latter setting to have a provably resistant aggregation method [67]. An alternative to requiring reliable reports is to restrict the percentage of malicious agents and their strategy space. For example, Dekel and Shamir [63] present an algorithm based on the support vector machine framework that can cope with maliciousness when the majority of agents provides accurate ratings, where accurate reports are described as samples from a common distribution. Contrary to the influence limiter algorithms, the above approaches do not consider the role of incentives in information gathering.

CHAPTER 8
Decentralized Machine Learning

The principles of data elicitation we have seen in the previous chapters need to be integrated in a complete system for gathering data to learn a model. This involves consideration of how the information agents are found and how they interact with the center, and how the center aggregates the data it receives into a model. In this chapter, we discuss issues that present themselves when designing the complete system so that the assumptions of the incentive mechanisms are satisfied.

We consider in particular the following issues.

- Selection and self-selection of data providers. The incentive techniques we have seen also drive a self-selection of data providers, since only those that have valid data can expect a net reward. However, the center will also want to select what data it wants to obtain, to avoid paying for redundant or uninteresting data. A related issue is ensuring that the community of reporting agents satisfies the assumptions of the game-theoretic mechanism. Depending on the domain, there may also be privacy issues for the data providers.
- The incentives derived from the game-theoretic analysis have to be translated into practical payment schemes. They have to be understandable and predictable to influence their behavior, and they have to be scaled so that they compensate their cost. Often, it is also not feasible to collect negative payments.
- Use of the data in machine learning or modeling techniques. Often, the model can be used to provide more robust incentives, and the incentives can be aligned with the loss function of the machine learning algorithm.

8.1 MANAGING THE INFORMATION AGENTS

An easy way to ensure the quality of data is to carefully select data providers, ideally to collect all the data oneself. The techniques we presented in this book enable a different kind of technique, that of *self-selection* where data providers volunteer to participate in the mechanism. For this self-selection to work, it is important that the mechanism should offer no rewards to agents whose data provides no information, and maximum rewards to those who provide the most valuable information.

What we want to eliminate are data that can be derived without observing the phenomenon. This includes random reporting according to a prior distribution, always reporting the same value, or any scheme where agents coordinate the data they provide based on a signal that is different from the phenomenon we want to observe. Many of the schemes described in this book are able to reduce the expected revenue of uninformative strategies to zero, and agents that intend to follow such strategies have no interest in participating in the platform. However, this always assumes *uncoordinated* strategies where each agent chooses its action for a particular task in isolation.

Group dynamics Most mechanisms assume that the agents play a single and non-repeated game. When used in a repeated setting, agents may learn to play a coarse correlated equilibrium that can be very different from the focal equilibria of the incentive mechanism. For example, Gao et al. [73] report on experiments where peer prediction mechanisms are applied in a repeated game setting, and agents learn to report according to an uninformative equilibrium.

In a repeated game setting, it may be more suitable to analyze behavior of incentive mechanisms in terms of *replicator dynamics*, where agents learn to choose their strategy according to the payoff observed during earlier instances of the game. Such an analysis is reported in Shnayder et al. [74] and shows that while output agreement and peer prediction are indeed vulnerable to uninformative equilibria, the collaborative agreement and peer truth serum schemes converge to the desired truthful equilibrium for a wide range of initial conditions.

Care must be taken to avoid collusive strategies where agents are able to systematically coordinate with their peer agents. For example, if all reporting agents determined their answers by applying a hash function to a task description, to the center it would be indistinguishable from honest reports in spite of providing no information.

One approach to counter coordinated low-quality strategies is to use *trusted agents* that provide the correct answers for some randomly selected subset of tasks. In a hybrid mechanism, agents' reports will be either compared to other agents, or to such trusted reports [69]. If the probability of having a trusted agents as a peer is sufficiently high, other low-quality equilibria can be broken. However, it has recently been shown that if the coordinated low-quality strategy provides higher payoffs than the cooperative strategy (for example, because it involves no measurement noise), it may be better to use simple truth agreement rather than a combination with peer consistency mechanisms as a complement to the truthful reports [70].

Self-selection In many scenarios, information agents choose themselves to participate or not in the mechanism. Therefore, there is a *self-selection* both of the information agents and of the data they provide about the phenomenon. Self-selection can be influenced by the incentives an agent can expected from participating. We would like, in particular, to *encourage* agents to provide data where the true value is very different from the prior, and to do so at high precision. On the other hand, we would like to *discourage* agents that provide data that is inaccurate or already known.

Table 8.1 shows the expected payments for the different mechanisms for an agent that had a prior expectation P and through measurement has obtained a posterior Q. We can see that the payment in general increases with the difference between the prior and posterior probabilities, and thus provides incentives to focus on data at uncertain locations. The only exception are schemes based on output agreement with a constant reward. However, here such an influence can be created by the center by making the constant depend on the prior uncertainty of the data.

Table 8.1: Comparison of expected payment from an agent's perspective for different mechanisms. The formulas are derived in the corresponding chapters. All mechanisms assume scaling so that answers according to the prior carry no reward. $H(P) = -\sum_x p(x) \log p(x)$ (Shannon Entropy), $\lambda(P) = \sum_x p(x)^2$ (Simpson's diversity index), and $\gamma(x) = q(x)/p(x) - 1$ (Confidence).

Mechanism	Expected Payment	Novelty	Precision
Truth Matching (value)	$max_x q(x) - max_x p(x)$	0 vs. 0	0.4 vs. 0.4
Truth Matching (log rule)	$H(P) - H(Q)$	0 vs. 0	0.648 vs. 0.728
Truth Matching (quadratic rule)	$\lambda(Q) - \lambda(P)$	0 vs. 0	0.32 vs. 0.28
Output Agreement	$max_x q(x) - max_x p(x)$	0 vs. 0	0.4 vs. 0.4
Peer Prediction (log rule)	$H(P) - H(Q)$	0 vs. 0	0.648 vs. 0.728
Peer Prediction (quadratic rule)	$\lambda(Q) - \lambda(P)$	0 vs. 0	0.32 vs. 0.28
Peer Truth Serum	$max_x \gamma(x)$	0 vs. 7	1 vs. 1.33
Correlated Agreement	$max_x[q(x) - p(x)]$	0 vs. 0.7	0.4 vs. 0.4
PTS for Crowdsourcing	$max_x \gamma(x)$	0 vs. 7	1 vs. 1.33
Logarithmic PTS	$D_{KL}(Q\|\|P)$	0 vs. 2.1	0.483 vs. 0.492
Bayesian Truth Serum	$D_{KL}(Q\|\|P)$	0 vs. 2.1	0.483 vs. 0.492
Divergence-based BTS (log)	$H(P) - H(Q)$	0 vs. 0	0.195 vs. 0.221
Divergence-based BTS (quadratic)	$\lambda(Q) - \lambda(P)$	0 vs. 0	0.32 vs. 0.28

While all schemes provide an incentive to report uncertain data, they do so to a different degree, as we will discuss in more detail below. The second goal is to encourage high precision. Here, the schemes could vary considerably.

To get an idea of how the incentives differ, we consider the following example scenarios. To evaluate the incentives for *novelty*—measuring at locations with frequently changing data—we compare the incentives for two points with three values that do not change to the same setup where the posterior indicates a different value from the prior and the probability distribution is permuted:

$$P_1 = (0.1, 0.8, 0.1), Q_1 = P_1 \text{ vs. } P_2 = (0.1, 0.8, 0.1), Q_2 = (0.8, 0.1, 0.1).$$

To evaluate incentives for measuring at higher *precision*, we compare a scenario with lower precision (three values) to one with higher precision (five values):

$$P_3 = (0.3, 0.4, 0.3), \qquad Q_3 = (0.1, 0.8, 0.1)\text{vs.}$$
$$P_4 = (0.1, 0.2, 0.4, 0.2, 0.1) \qquad Q_4 = (0.05, 0.1, 0.7, 0.1, 0.05).$$

The results are shown in Table 8.1 in the columns labeled *Novelty* and *Precision*. We see big difference between the schemes.

For the *novelty* scenario, all of the mechanisms have an expected reward of zero when the value does not change—this is the result of normalization. However, many of the mechanisms provide no incentive for measuring changing values at all! This is because these mechanisms compensate for reporting according to the prior through a constant offset that does not depend on the reported value. Thus, the expected reward only depends on the shape of the distribution, but not on the actual values. An agent that reports a different value, but with a posterior that has the same shape as the prior as in the novely scenario, thus gets the same reward as when it reports according to the prior—nothing.

In contrast, in the PTS, CA, and BTS mechanisms the compensation is dependent on the reported value, and they can thus distinguish a noisy observation where the value has changed from one where it has not. Note, however, that when the reported data has a higher certainty than the prior, all schemes yield a positive expected reward and thus do encourage participation.

For the *precision* scenario, we find that some mechanisms neither encourage or discourage improved precision: they are truth matching, output agreement and correlated agreement. All peer truth serums and all mechanisms based on the logarithmic scoring rule do encourage higher granularity. However, mechanisms based on the quadratic scoring rule actually discourage it!

These are only example scenarios, and we do not have a general analysis as we do not have a good way of classifying scenarios. However, we have observed these effects empirically as well. In the pollution sensing example, Figure 3.10 indeed shows that the Peer Truth Serum has a stronger tendency to incentivize measurement at uncertain locations than other mechanisms.

Other issues An important issue is whether agents know of each other's reports. If they do, it opens up possibilities for collusion, but at the same time it also lets agents have more homogeneous beliefs that will better fit the mechanism. Some mechanisms, such as BTS, assume that agents have no knowledge of other reports, while others such as PTS allow publication of partial results as long as agents have no knowledge of current peer reports. Which situation is appropriate will generally be dictated by the application.

In sensing applications, such as pollution sensing, there is often a goal of selecting a minimum number of sensors to provide full coverage of the space. However, peer mechanisms require a certain redundancy in order to validate data, which is often in contradiction with minimality.

An issue related to precision is the risk of agents coordinating on a signal other than the phenomenon itself, called low-quality signal in Gao et al. [70]. Such signals are typically less uncertain and might have fewer values than the phenomenon. A mechanism that does not sufficiently incentivize precision may encourage agents to report such low-quality signals instead, especially since they are often easier to detect.

Finally, a potential issue is that agents might hold back information in the hope of being rewarded more by showing it later, a strategy that is called *counterspeculation*. This can be very damaging to the actual performance, and mechanisms have to be carefully designed to discourage such behavior. One way to do this is to give rewards based on myopic impact, i.e., the impact that a report has on the quality of information held by the center at the precise time the data is received. Fortunately, such myopic rewards are often also the easiest to implement, but they can leave open the door to sophisticated manipulations that repeatedly report different data with the goal of collecting multiple myopic rewards, and that have to be ruled out by restricting the possibility of the same agent providing multiple reports over time.

8.2 FROM INCENTIVES TO PAYMENTS

The schemes we discussed in this book are designed to ensure that agents adopt cooperative, truthful strategies. As we discussed in Chapter 2, generally the values given by the incentive scheme need to be transformed into payments by proper scaling. In fact, any positively monotone transformation is admissible while leaving the incentive properties intact.

The simplest approach is to scale an incentive *inc* by a linear transformation $pay = \alpha(inc + \beta)$, where $\alpha > 0$ is chosen so that the difference between truthful and non-truthful reports covers the effort of observation, and β ensures that the expected payment for uninformed reporting is equal to zero.

However, there are two drawbacks:

1. it may require payments that are sometimes negative, and
2. because of measurement noise, the payments can be very volatile.

Figure 8.1 illustrates the volatility of payments for two different scenarios: a simulated crowdsourcing scenario (left) and a crowdsensing scenario using the data from Hasenfratz et al. [71]. For the crowdsourcing scenario, we assume the payment scheme of Dasgupta and Ghosh [41], while the crowdsensing scenario scores against the ground truth using a quadratic scoring rule (as described in Chapter 2). As we can see, both schemes result in very volatile payments where the difference between accurate and random reports is much smaller than the payments themselves.

If we wanted to make the expected payment for random reporting equal to zero by subtracting a proper β, payments would very often be negative and appear even more volatile.

When agents are free to continuously revise their decisions on their strategy, it is important to make the payments more predictable. When the same agent interacts repeatedly with the center, this can be achieved by smoothing the payments over multiple interactions, so that negative and positive payments balance out and we obtain a more stable positive reward.

To implement this idea, we propose a reputation mechanism similar to the influence limiter we presented in Chapter 6. It can use any of the incentive mechanisms we presented in this

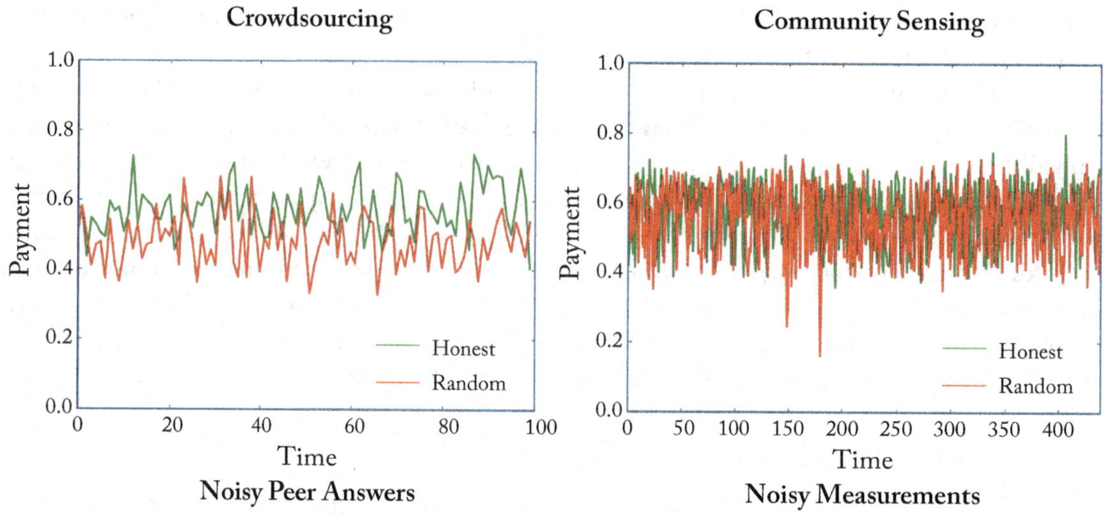

Figure 8.1: Variability of payments for accurate vs. inaccurate answers in crowdsourcing and community sensing (based on [72]).

book to generate a *score* that identifies the quality of the data reported by the agent. However, instead of making this the basis for an immediate payment, we use it to update the *reputation* of the agent. The reputation in turn determines the payments that the agent can get on future tasks.

For the agent, this creates a clear situation where it always knows what payment it will get for the *next* data it reports to the center. Thus, the payment an agent will receive is always completely predictable, and the agent can decide if it is sufficient to compensate the cost of observing the phenomenon.

Furthermore, the use of reputation acts to smooth the payments. Where the incentive scheme would prescribe negative payments, we now simply decrease the reputation and thus future payments.

Different from the influence limiter we presented in Chapter 6, we do not measure the impact of the report on the quality of the learned model, but score it by agreement with a peer, for example according to the PTS scheme. The peer can also be the prediction $\hat{\theta}_{\mathcal{F}}$ by a model $\mathcal{F}$ on the basis of multiple peer reports. To obtain linearity, here we propose to use the version derived from the quadratic scoring rule (see Chapter 3):

$$\pi_t(x) = \mathbf{1}_{\hat{\theta}_{\mathcal{F}}=x} - \Pr(\hat{\theta}_{\mathcal{F}} = x)$$
$$score_t(x) = (1-\alpha) \cdot \pi_t(x) - \alpha.$$

The prior probability $Pr(\hat{\theta}_{\mathcal{F}} = x)$ can be estimated using the model and prior data. The parameter α determines the minimal acceptable quality, i.e., the threshold below which reputation should decrease.

Based on this score, in the PropeRBoost scheme [72] we maintain the reputation rep_t of an agent at each time t using a similar framework as in the influence limiter.

- We scale positive payments by $\sigma = \frac{rep_t}{rep_t+1}$, and this scaling factor is communicated to the agent before its report.
- The agent picks a task, submits the data and gets paid $\sigma \cdot \tau$, where τ is a payment function.
- The mechanism determines the score $score_t$ of the submitted data, and updates the reputation using the following exponential updating rule:

$$rep_t = rep_t \cdot (1 + \eta \cdot score_t),$$

  where $\eta \in (0, \frac{1}{2}]$ is a learning parameter.

PropeRBoost has the following properties [72]:

- The average payments to accurate agents are near maximal.
- The average payments for agents who consistently have low proficiency are near minimal.
- The average payments to agents whose proficiency converges toward levels lower than p_l are near minimal.

Together these properties imply a much less volatile separation between incentives for good and low-quality work, as we can also observe in empirical evaluations.

Figure 8.2 shows the performance of the scheme in the two simulated scenarios. We can see that it creates a very strong separation of payments between cooperative and random strategies, as compared with the "raw" incentives shown in Figure 8.1. Also note that there is no need for negative payments.

Figure 8.3 shows the evolution of the scaling factor σ for four different strategies.

1. *Random*: report randomly according to the prior distribution.
2. *Honest*: cooperative strategy.
3. *Switch*: cooperative strategy for the first half, then random.
4. *KeepRep*: cooperative strategy whenever scale falls below threshold, random reports when above.

We can see that in both scenarios the reputation system keeps track of the correct reputation quite accurately: random reports soon obtain no payment at all, while cooperative strategies converge toward the maximum scale.

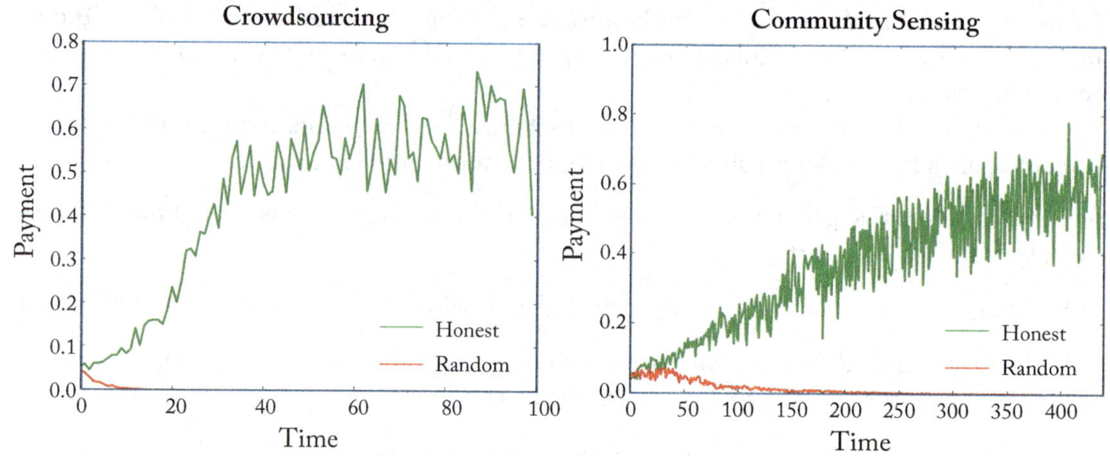

Figure 8.2: Paymenst using the PropeRBoost scheme in crowdsourcing and community sensing scenarios, for honest and random reporting (based on [72]).

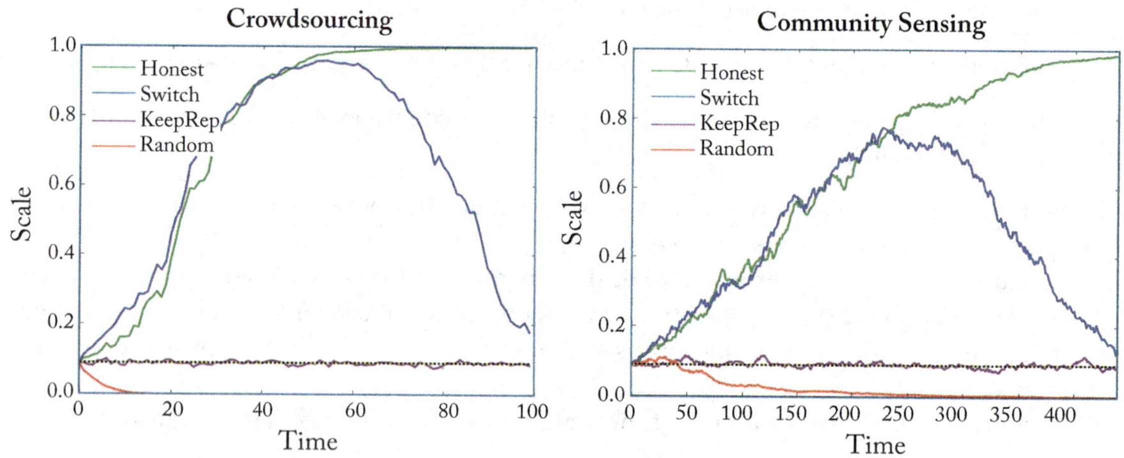

Figure 8.3: Scaling factors of the PropeRBoost scheme in crowdsourcing and community sensing scenarios, for four different strategies (based on [72]).

8.3 INTEGRATION WITH MACHINE LEARNING ALGORITHMS

An important consideration is how the data will be used. Most often, it will be processed by some machine learning algorithm to obtain a model, and the real goal is the quality of the model.

There are many machine learning algorithms and the exact analysis of the impact of information gathering on the learning outcome will of course depend on the details of the algorithm used.

The techniques we have seen generally apply to elicitation of a value out of a finite set of possibilities, or *classification* models. The machine learning algorithm learns a function $f(Z) = \Pr(X|Z)$ that gives an estimate of the class X given the features Z.

The agents provide data for different features z, and this data is used by the machine learning algorithm to compute its model. The data should be chosen to optimize the convergence of the machine learning algorithm. Among the many possibilities, we consider two ways this could be done.

- *Histogram*: The center collects data for the same parameter set z from multiple peer agents, and the goal is to obtain a hstogram that approximates the true probability distribution of x for these features as closely as possible. We measure the distance by a logarithmic loss criterion, in particular the Kullback-Leibler Divergence. This applies very well for example to product reviews, where different reviewers evaluate exactly the same product.

- *Interpolation*: The center integrates all received data into its model, and uses the prediction of the model as the peer estimate. Here the data does not necessarily have the same features z, but the peer report could be for different features z' where the model allows interpolation between the features. The collected data should support convergence of the model as well as possible. We measure the convergence by the Brier score of the model against a peer report. This applies well, for example, to pollution measurements, where every measurement is taken at a slighly different point, and the points are interpolated by the model.

- *Classification*: Recently, incentive schemes have also been shown for scenarios where the center learns a classifier from labels provided by information agents.

8.3.1 MYOPIC INFLUENCE

We can generalize the idea of incentivizing reports of data that maximally improves a learning result to other learning algorithms. In the general formulation, we should reward agents for the immediate influence their data has on the quality of the learned model. We call this their *myopic influence* since it only considers the current step. We have used this notion in the chapter on the stochastic influence limiter.

For many reward schemes, it is straightforward to relate the reward to the myopic influence of a report. While such a greedy approach cannot guarantee that data providers provide the optimal *combination* of data for the learner, it is the best that can be done when rewarding individual data items.

In almost all cases, basing rewards on myopic influence has another beneficial effect: since there tends to be diminishing return of data on the model quality, an agent would not want to speculate on getting a higher reward by holding back a report of its data until a later time.

8.3.2 BAYESIAN AGGREGATION INTO A HISTOGRAM

First, we consider the simplest case where the data collected are discrete values, and the learning algorithm aggregates them in a Bayesian way, i.e., using Equation (1.1), to form a histogram that increasingly approximates the true probability distribution. This covers many of the scenarios mentioned in the introduction, such as aggregating review scores or pollution measurements.

More precisely, consider that the center updates a normalized histogram $R^t = (r^t(x_1), \ldots, r^t(x_n))$ at time t. When receiving report of a value x_i, it sets:

$$\begin{aligned} r^{t+1}(x_i) &= \frac{t r^t(x_i) + 1}{t+1} \\ r^{t+1}(x_j) &= \frac{t}{t+1} r^t(x_j). \end{aligned}$$

The most common measure to evaluate the quality of approximating a distribution Q by the estimate R is the Kullback-Leibler divergence, expressed as:

$$D_{KL}(R||Q) = \sum_{i=1}^{n} q(x_i)(\ln q(x_i) - \ln r(x_i)) = -H(Q) - E_Q[\ln r(x_i)].$$

The divergence is minimized by minimizing the second term, $E_Q[\ln r(x_i)]$. Note that this is just the expected score for a randomly chosen observation o obtained by the logarithmic scoring rule, i.e.:

$$E_Q[\ln r(x_i)] = E[\ln r(o)] = E[\ln r(x_p)],$$

where the second inequality holds provided that the agent believes peer agents to report truthfully. This is exactly the reward scheme that rewards agents according to the logarithmic scoring rule applied to a randomly chosen peer report, such as the peer prediction method and the peer truth serum. Thus, as long as agents believe that peer reports are accurate measurements of the phenomenon, the scheme makes agents report the values that also result in the best convergence of the results learned by the center!

Helpful reporting in the peer truth serum We can apply this observation to analyze the properties obtained by the helpful reporting strategies (Definition 3.7) we discussed for the Peer Truth Serum in Section 3.3.2. Assuming Bayesian aggregation as detailed above, we can show that as long as agents adopt helpful strategies, the distribution R will converge to the true distribution P^* [20]. We call this property *asymptotically accurate*.

Definition 8.1 A mechanism for information elicitation is *asymptotically accurate* if it admits an equilibrium such that the averaged reports converge to the true distribution of the phenomenon, for a stationary phenomenon. And we can show the following [20].

Theorem 8.2 *The Peer Truth Serum with informed prior beliefs is asymptotically accurate, provided that agents' belief updates satisfy the self-predicting condition.*

As the Peer Truth Serum provides rewards according to the logarithmic scoring rule, it implements the learning algorithm discussed above. We can thus see that untruthful but helpful reports are actually maximize the speed of convergence of the overall learning system to be faster than truthful reporing alone!

Similarly, if we wanted to optimize the Brier score of the histogram, i.e., minimize the mean squared error of R with respect to S, we obtain the analogous result for schemes that use the quadratic scoring rule.

8.3.3 INTERPOLATION BY A MODEL

When data is used to learn a model that can be used for interpolation, it is not even necessary to have an exact peer. Instead, we can evaluate the improvement of the model obtained by the agent's report by comparing it to a randomly selected peer report. Letting f_{-i} be the model without the data from agent i, and f the model that incorporates the data. To compute the reward for agent i's data, we compute the difference in evaluation that the model provides for the features z' of a randomly chosen peer report, i.e., $SR(f(z'), x_p) - SR(f_{-i}(z'), x_p)$.

Ideally, we take the same approach as in the derivation of the peer truth serum, i.e., we approximate the difference by a first-order Taylor expansion around the current model f_{-i}, and make the reward proportional to the derivative. However, the difficulty is that we need to know the derivative of the learning algorithm with respect to the new example, for the features associated with the peer report. Furthermore, since the peer report is selected randomly, the mechanism cannot assume a particular set of features z' for the derivative, but only an expectation. When the model is an interpolation model, the shadowing functions used in the derivation of the peer truth serum are reasonable to use in expectation. We show an evaluation of such a model in the pollution example in Chapter 5.

Another possibility is to let the model interpolate between neighboring peer reports, and thus to generate an artificial peer report as the model prediction for the feature set z that is identical to that encountered by agent i. While it would seem that this encourages agents to report according to the already existing model, it is actually just the converse of the approach in the previous paragraph: rather than using a Taylor approximation around the features z, we use an approximation around the set z', and so the effect can be expected to be the same. However, it is much easier to implement. We show an evaluation using their approach where we use the basic PTS for the pollution example, in Chapter 3.

8.3.4 LEARNING A CLASSIFIER

For a setting with binary signals where the center learns a classifier, Liu and Chen [77] show a learning technique such that peer prediction using the prediction of a learned classifier as peer report provably has truthful reporting as the highest-paying equilibrium. The main difficulty with guaranteeing truthfulness is that the data provided by peers may itself be biased, and thus incentives are given for matching a biased version of the truth.

To avoid this issue, the bias in the classifier is removed based on the the average classification error rates, or flipping rates, which are assumed uniform among all workers. They are characterized by separate success rates p_+ for positive examples and p_- for negative examples. Provided that the true probability of an example being positive $\mathcal{P}_+$ is known, they can be found from the observed probability of positive labels p^+ and the observed average rate of agreement of two agents on the same tasks q, by solving the system of equations:

$$\begin{aligned} q &= \mathcal{P}_+[p_+^2 + (1-p_+)^2] + (1-\mathcal{P}_+)[p_-^2 + (1-p_-)^2] \\ p^+ &= \mathcal{P}_+ p_+ + (1-\mathcal{P}_+)p_-. \end{aligned}$$

As shown in Liu and Chen [77], knowing these success rates allows to learn an unbiased classifier that in turn can be used as a basis for an incentive scheme, and the unbiasing can be integrated into the payment function. Even if the computation of the p_+ and p_- requires a set of tasks that are solved by multiple agents in order to determine the agreement rate, this set can be a tiny fraction and for all other tasks only a single report is required.

The approach is restricted to binary signals and homogeneous agent populations, and it is difficult to extend to more complex cases while maintaining the possibility to prove its incentive properties.

8.3.5 PRIVACY PROTECTION

In some cases, information agents may require that the mechanism protects the privacy of their data. This can be achieved when the data is aggregated into a model, such as using a machine learning algorithm. Such a scheme is described in Waggoner et al. [75], where rewards are given for improvement of the performance of the learned model on test data. Ghosh et al. [76] give a more precise analysis using the framework of differential privacy. The privacy protection given by these mechanisms depends crucially on the fact that data is aggregated into a model where individual contributions are consequently hard to determine.

8.3.6 RESTRICTIONS ON AGENT BEHAVIOR

In some cases, it may be possible to place additional assumptions on worker behavior, so that only certain randomized reporting strategies are allowed. For example, Cai et al. [78] consider a setting where agents observe objective values and their reports are drawn from a distribution centered around the true value with some Gaussian noise. The only parameter of the reporting strategy is the amount of effort which is inversely proportional to the noise. Even if this setting is more restrictive, it can model many real situations, including many sensor networks, although it is inapplicable to a situation where sensors do not measure at all, or their average measurement is not equal to the true value.

For this setting, they propose a simple mechanism that pays a report x_i of agent i according to

$$c_i - d_i(x_i - \hat{f}_{-i})^2,$$

where $\hat{f}_{-i}$ is the model constructed without the report from agent i. Their result is impressive in two aspects: first, it applies to a wide range of machine learning models for the model $\hat{f}$, including many forms of regression. Second, exerting maximum effort is a *dominant* strategy, not just a Nash equilibrium—no matter what the other agents do, exerting maximum effort is *always* the best response. The strong assumptions that are required do not make this a generally applicable solution, but it shows that there may be good possibilities.

CHAPTER 9
Conclusions

Information systems are increasingly based on large collections of data, often obtained from multiple sources outside the direct control of the system designer. Since obtaining accurate data is costly, data providers need to be compensated for their effort. This requires that we can evaluate the quality of the data so that only accurate data is paid.

In this book, we presented a variety of game-theoretic mechanisms that incentivize participants to provide truthful data, while penalizing those that provide poor data and discouraging them from participating. This field of research is still in its infancy, and our overview can only be seen as a snapshot of the state of the art at this time. We hope that it will help to spread understanding of the techniques that are known today, and foster further development of the field. We also believe that recent progress has given us mechanisms that allow for the first time to pay for data according to its quality, an important milestone for the development of data science.

We started by noting that there are three different ways of controlling the quality of data. The first and most well known one is to filter outliers using statistical techniques. The second is to learn the average quality of data provided by each agent, and assume that data will be of the same quality in the future. One of example of such an approach is the influence limiter reputation scheme we described in Chapter 7.

However, most of this book has been devoted to the third option, which is to provide *incentives* that make information agents do a better job at providing relevant and accurate data. They are a key because they actually increase the amount of high-quality data that is available in the first place.

9.1 INCENTIVES FOR QUALITY

We first started with the incentive mechanism design and described two most important classes of incentives: those based on a known ground truth, and those that are based on comparison of reported values, i.e., peer consistency methods.

The main advantage of mechanisms based on the ground truth is that they make cooperative strategies not just an equilibrium, but a *dominant* strategy—it is best independently of what other information agents do. Thus, whenever the ground truth is available, it is preferable to take advantage of it to obtain these stronger incentive properties.

However, most contributed data cannot be validated on a ground truth, either because this is too costly, or because it is subjective. Peer consistency methods therefore make up the bulk of the techniques in this book.

When the center has no independent verification of the reported data at all, it is of course possible for the information agents to fake an entirely different reality to the center, by all agreeing to report data that may not even be related to the information that is requested. Clearly, this possibility can only be ruled out by having some access to ground truth, for example through spot checks. On the other hand, it is unlikely that such a coordinated deception would not be detectable by the center through other means.

9.2 CLASSIFYING PEER CONSISTENCY MECHANISMS

We have presented a large variety of peer consistency mechanisms that can be applied to different scenarios. The characteristics of the application will impose specific requirements that should guide the choice of the best mechanism. To enable the right selection, we provide a classification according to five different criteria that determine their fit with an application.

Size of the task space (Figure 9.1) The first criterion has to do with how much data is being elicited from each information agent. Some mechanisms, such as peer prediction and BTS, apply even when only a single data item is elicited. Others, such as Log-PTS and PTSC, require that the agent population answers to a larger number of (a priori) similar elicitation tasks to provide a stronger mechanism. In between, we have correlated agreement that does require multiple similar tasks, but can do with a small number as long as the correlations are known beforehand.

In general, the application will dictate what assumption can be made. Opinion polls and reviews will generally not have multiple similar tasks, while crowdsourcing and peer grading usually do.

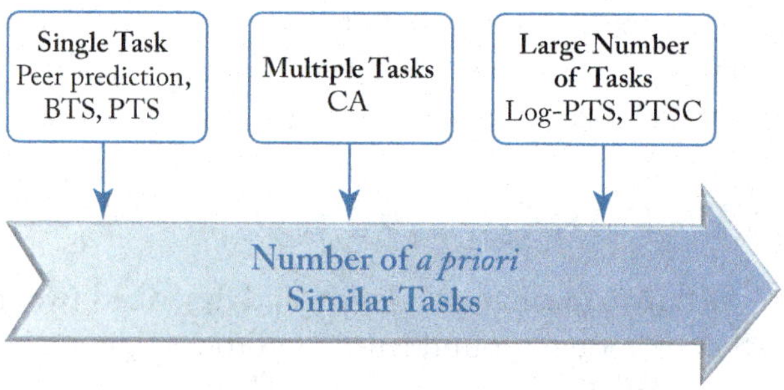

Figure 9.1: Amount of information elicited per task.

Information elicited from each agent (Figure 9.2) Some mechanisms require agents to only provide the data, while others such as BTS also require additional prediction reports (which may be more voluminous than the data itself). Some mechanisms, such as CA, require that agents submit multiple data items for similar tasks.

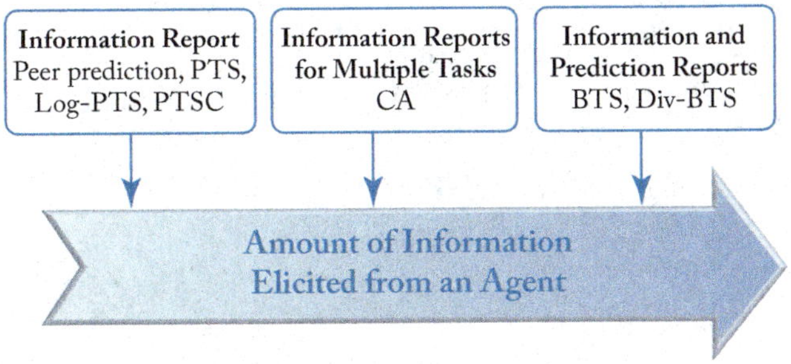

Figure 9.2: Amount of information elicited per task.

Number of peers required (Figure 9.3) Peer consistency relies on comparing reports of peers on the same (or at least highly similar) tasks. In some applications, such as product reviews, there are always many peers that answer to exactly the same task, so we can apply mechanisms such as BTS that require a large population of such peers. On the other hand, in crowdwork having peers solve the same task is a waste of effort, so we want to minimize it and use mechanisms such as peer prediction (which could even use peers that solve tasks with only statistically correlated answers). As a third possibility, we have PTSC, which requires a larger number of agents to solve many tasks, but only one peer per task.

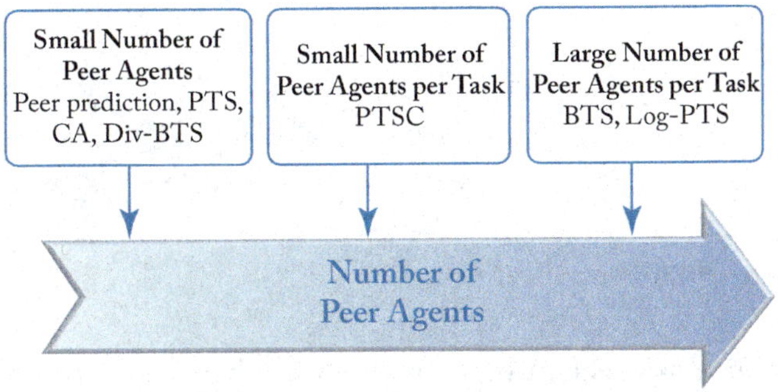

Figure 9.3: Number of peers required by the mechanism.

Knowledge of agent beliefs (Figure 9.4) Some mechanisms are tuned closely to agents' belief structures, and their design requires to know these beliefs quite precisely. A big requirement for applying peer prediction mechanisms is to know agents' posterior beliefs for different observed signals. PTS, on the other hand, requires knowledge of prior beliefs, which are often easy to obtain and do not depend on observed signals. The big advantage of BTS mechanisms is that they obtain knowledge about the beliefs through the additional prediction reports, and thus require little knowledge by the mechanism designer.

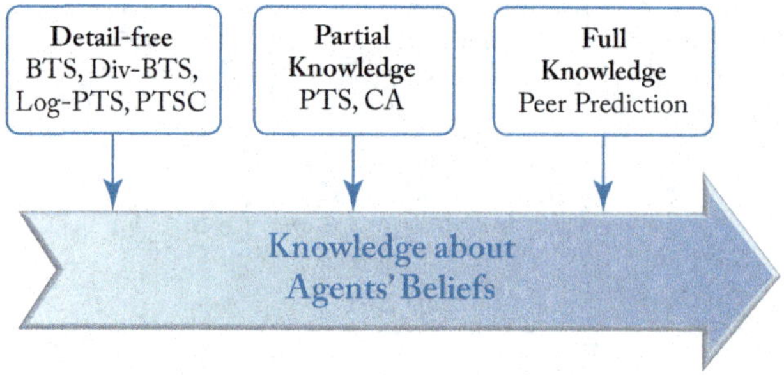

Figure 9.4: Information required about agent beliefs.

Strength of incentives (Figure 9.5) All of the mechanisms we have seen have strict equilibrium strategies that are truthful and (with the right scaling) cooperative. However, there are differences in the strength of these equilibria. In peer prediction, the truthful equilibrium usually does not have the highest expected reward, and it requires specific care in its design to make the truthful equilibrium focal.

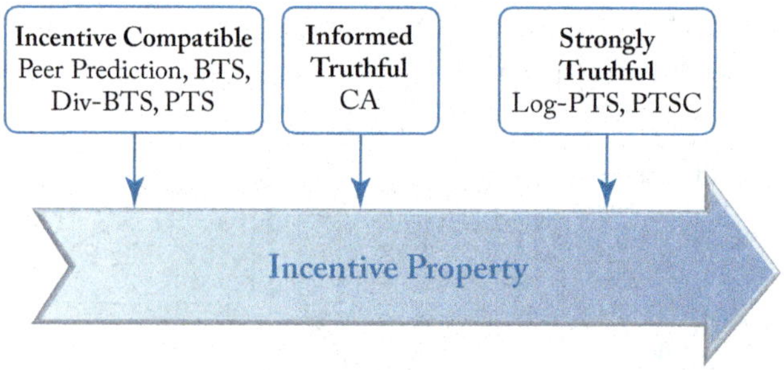

Figure 9.5: Relative strength of the incentive.

Most of the multi-task mechanisms, on the other hand, are strongly truthful, meaning that they guarantee that truthful reporting is not only a strict equilibrium, but also the highest paying one.

Correlated Agreement also guarantees that truthful reporting results in a greater payoff than any other uninformed reporting strategy (strategies for which an agent does not make an observation). However, it provides no incentive for distinguishing correlated values.

We have further seen in Chapter 8 that there are big differences in the expected rewards of the different schemes, depending on the elicitation scenario. These expected rewards have an influence on the motivation of information agents to participate in the mechanisms, and thus the population of information agents that will be formed by self-selection.

9.3 INFORMATION AGGREGATION

It would be nice to also incentivize agents to participate in an optimal aggregation of information, as here also information agents are likely to have better knowledge than the center. We discussed two schemes where incentives also involve information aggregation: prediction markets and reputation systems inspired by the prediction market framework. We addressed two questions:

- how to elicit agents' confidence together with the desired information, and
- how to limit the negative impact of the malicious agents on the learned outcome.

The first one can be addressed by a prediction market, designed analogously to stock markets to implicitly elicit confidence that agents have in their predictions. The second one, the influence limiter, is designed specifically for online information fusion where information providers have to be kept from manipulating the aggregated result. Both methods use a verifiable source of information to determine the quality of agents' reports.

Other work in a similar direction that we did not discuss include wagering schemes that make agents reveal their confidence in a prediction, and consensus schemes such as the Delphi method where agents iteratively refine their report to reach a consensus. We did not discuss these schemes as they involve more complex strategies than simply reporting true data, which should not be negotiable.

9.4 FUTURE WORK

There are many directions that one can take in extending the mechanisms presented in this book. Some open issues that seem important to us are as follows.

- Most work on peer consistency techniques assumes that an agent and its peer observe the exact same underlying phenomenon. This is often unrealistic: we will not have multiple sensor measurements at the same location, and it is wasteful to give the same task to multiple crowdworkers. Experimentally, many of the mechanisms we described work well using signals for highly correlated rather than the same measurements. However, there are no proofs that these are indeed truthful equilibria.
- When data involves a large range of possible answers, the differentiation between signal values becomes very small, and incentives thus become increasingly volatile. For the extreme case of continuous values, we have seen that is possible to extend the divergence-based BTS mechanism. We conjecture that it is also possible to extend the PTS mechanism using a similar construction of randomly chosen intervals, but so far no such extension has been worked out.
- When information agents are not rational, or even malicious, it is desirable to bound their influence on the outcome. The influence limiter method we presented achieves this, but

only when information can be verified. It would be interesting to see under what condition it is possible to extend the discussed approaches to allow peer consistency based evaluation where only a fraction of agents can be trusted.

- When agents have the goal of influencing the result of a machine learning algorithm that will be applied to the data they provide, the learning algorithm will have an influence on the incentives. The interplay between incentives and specific machine learning algorithms is complex to analyze because the influence of machine learning algorithms is often highly nonlinear.

Bibliography

[1] Nan Hu, Paul A. Pavlou, and Jennifer Zhang, Can online reviews reveal a product's true quality? Empirical findings and analytical modeling of Online word-of-mouth communication, *7th ACM Conference on Electronic Commerce*, pp. 324–330, 2006. DOI: 10.1145/1134707.1134743. 2

[2] Nan Hu, Paul A. Pavlou, and Jennifer Zhang, On self-selection bias in online product reviews, *Management Information Systems Quarterly*, 2017. DOI: 10.25300/misq/2017/41.2.06. 2

[3] Vikas C. Raykar, Shipeng Yu, Linda H. Zhao, Gerardo Hermosillo Valadez, Charles Florin, Luca Bogoni, and Linda Moy, Learning from crowds, *Journal of Machine Learning Research*, **11**, pp. 1297–1322, 2010. 6

[4] Yuqing Kong and Grant Schoenebeck, A framework for designing information elicitation mechanisms that reward truth-telling, *arXiv:1605.01021*, 2016. 10, 66, 74, 81

[5] Alexander P. Dawid and A. M. Skene, Maximum likelihood estimation of observer error-rates using the EM algorithm, *Journal of the Royal Statistical Society. Series C (Applied Statistics)*, **28**(1), pp. 20–28, 1979. DOI: 10.2307/2346806. 6, 13, 75

[6] Nicolas Lambert, David M. Pennock, and Yoav Shoham, Eliciting properties of probability distributions, *Proc. of the 9th ACM Conference on Electronic Commerce (EC)*, pp. 129–138, 2008. DOI: 10.1145/1386790.1386813. 22

[7] Nicolas Lambert and Yoav Shoham, Eliciting truthful answers to multiple-choice questions, *Proc. of the 10th ACM Conference on Electronic Commerce (EC)*, pp. 109–118, 2009. DOI: 10.1145/1566374.1566391. 22

[8] Luca de Alfaro, Marco Faella, Vassilis Polychronopoulos, and Michael Shavlovsky, Incentives for truthful evaluations, *arXiv:1608.07886*, 2016. 23

[9] Glenn W. Brier, Verification of forecasts expressed in terms of probability, *Monthly Weather Review*, **78**(13), 1950. DOI: 10.1175/1520-0493(1950)078%3C0001:vofeit%3E2.0.co;2. 24

[10] Irving John Good, Rational decisions, *Journal of the Royal Statistical Society*, **14** pp. 107–114, 1952. DOI: 10.1007/978-1-4612-0919-5_24. 24

[11] Tilmann Gneiting and Adrian E. Raftery, Strictly proper scoring rules, prediction, and estimation, *JASA*, **102**(477), 2007. DOI: 10.21236/ada459827. 24

[12] Edward H. Simpson, Measurement of diversity, *Nature*, pp. 163–688, 1949. DOI: 10.1038/163688a0. 25

[13] Luis von Ahn and Laura Dabbish, Designing games with a purpose, *Communications of the ACM*, **51**, pp. 58–67, 2008. DOI: 10.1145/1378704.1378719. 29

[14] Yang Liu and Yiling Chen, Learning to incentivize: Eliciting effort via output agreement, *Proc. of the 25th International Joint Conference on Artificial Intelligence (IJCAI)*, 2016. 31

[15] Nolan Miller, Paul Resnick, and Richard Zeckhauser, Eliciting informative feedback: The peer prediction method, *Management Science*, 2005. DOI: 10.1287/mnsc.1050.0379. 32, 33

[16] Radu Jurca and Boi Faltings, Mechanisms for making crowds truthful, *JAIR*, **34**, pp. 209–253, 2009. 31, 36

[17] Yuqing Kong, Katrina Ligett, and Grant Schoenebeck, Putting peer prediction under the micro(economic)scope and making, *arXiv:1603.07319*, 2016. DOI: 10.1007/978-3-662-54110-4_18. 38

[18] Rafael Frongillo and Jens Witkowski, A geometric method to construct minimal, *Proc. of the 30th AAAI Conference on Artificial Intelligence*, pp. 502–508, 2016. 32, 39, 40, 41

[19] Jens Witkowski and David C. Parkes, A robust Bayesian truth serum for small populations, *AAAI*, 2012. 30, 41

[20] Radu Jurca and Boi Faltings, Incentives for answering hypothetical questions, *SCUGC*, 2011. 32, 47, 116

[21] Radu Jurca, Boi Faltings, and Walter Binder, Reliable QoS monitoring based on client feedback, *Proc. of the 16th International Conference on World Wide Web (WWW)*, pp. 1003–1012, 2007. DOI: 10.1145/1242572.1242708. 48

[22] Boi Faltings, Jason J. Li, and Radu Jurca, Incentive mechanisms for community sensing, *IEEE Transaction on Computers*, **63**(1), pp. 115–128, 2014. DOI: 10.1109/tc.2013.150. 45, 49, 50

[23] Gustav Theodor Fechner, Elements of psychophysics, *Breitkopf und Härtel*, 1860. DOI: 10.1037/11304-026. 55

[24] Jan Lorenz, Heiko Rauhut, Frank Schweitzer, and Dirk Helbing, How social influence can undermine the wisdom of crowd effect, *Proc. of the National Academy of Sciences*, **108**(22), pp. 9020–9025, 2011. DOI: 10.1073/pnas.1008636108. 57

[25] Boi Faltings, Pearl Pu, Bao Duy Tran, and Radu Jurca, Incentives to counter bias in human computation, *Proc. of HCOMP*, pp. 59–66, 2014. 57, 58

[26] Elaine E. Marconi, Experience of a Lifetime, 2007. `https://www.nasa.gov/mission_pages/station/behindscenes/student_visit.html` 57

[27] Shih-Wen Huang and Wai-Tat Fu, Enhancing reliability using peer consistency evaluation in human computation, *Proc. of the Conference on Computer Supported Cooperative Work*, pp. 639–648, 2013. DOI: 10.1145/2441776.2441847. 57, 82

[28] Drazen Prelec, A Bayesian truth serum for subjective data, *Science*, **306**(5695), pp. 462–466, 2004. DOI: 10.1126/science.1102081. 59, 61

[29] Drazen Prelec and Sebastian Seung, An algorithm that finds truth even if most people are wrong, Unpublished manuscript, 2006. 69

[30] Ray Weaver and Drazen Prelec, Creating truth-telling incentives with the Bayesian truth serum, *Journal of Marketing Research*, **50**(3), pp. 289–302, 2013. DOI: 10.1509/jmr.09.0039. 70

[31] Leslie K. John, George Loewenstein, and Drazen Prelec, Measuring the prevalence of questionable research practices with incentives for truth-telling, *Psychological Science*, **23**(5), pp. 524–532, 2012. DOI: 10.2139/ssrn.1996631. 70

[32] Drazen Prelec, H. Sebastian Seung, and John McCoy, A solution to the single-question crowd wisdom problem, *Nature*, **541**, pp. 532–535, 2017. DOI: 10.1038/nature21054. 70

[33] Jens Witkowski and David C. Parkes, A robust Bayesian truth serum for small populations, *Proc. of the 26th Conference on Artificial Intelligence (AAAI)*, 2012 62

[34] Goran Radanovic and Boi Faltings, A robust Bayesian truth serum for non-binary signals, *Proc. of the 27th Conference on Artificial Intelligence (AAAI)*, 2013. 32, 62, 63, 64

[35] Goran Radanovic and Boi Faltings, Incentives for truthful information elicitation of continuous signals, *Proc. of the 28th Conference on Artificial Intelligence (AAAI)*, 2014. 64, 66, 67

[36] Goran Radanovic, Elicitation and aggregation of crowd information, Ph.D. thesis (EPFL), 2016. 32, 64, 67

[37] Yuqing Kong and Grant Schoenebeck, Equilibrium selection in information elicitation without verification via information monotonicity, *arXiv:1603.07751*, 2016. 64, 66

[38] Jens Witkoswki and David C. Parkes, Peer prediction without a common prior, *Proc. of the 13th ACM Conference on Electronic Commerce (EC)*, pp. 964–981, 2012. DOI: 10.1145/2229012.2229085. 69

[39] Peter Zhang and Yiling Chen, Elicitability and knowledge-free elicitation with peer prediction, *Proc. of the 13th International Conference on Autonomous Agents and Multiagent Systems (AAMAS)*, pp. 245–252, 2014. 69

[40] Radu Jurca and Boi Faltings, Incentives for expressing opinions in online polls, *Proc. of the 9th ACM Conference on Electronic Commerce (EC)*, pp. 119–128, 2008. DOI: 10.1145/1386790.1386812.

[41] Anirban Dasgupta and Arpita Ghosh, A Crowdsourced judgement elicitation with endogenous proficiency, *Proc. of the 22nd International Conference on World Wide Web (WWW)*, pp. 319–330, 2013. DOI: 10.1145/2488388.2488417. 71, 72, 111

[42] Victor Shnayder, Arpit Agarwal, Rafael Frongillo, and David. C. Parkes, Informed truthfulness in multi-task peer prediction, *Proc. of the ACM Conference on Economics and Computation (EC)*, pp. 179–196, 2016. DOI: 10.1145/2940716.2940790. 72, 75

[43] Arpit Agarwal, Debmalya Mandal, David Parkes, and Nisarg Shah, Peer prediction with heterogeneous users, *Proc. of the ACM Conference on Economics and Computation (EC)*, pp. 81–98, 2017. DOI: 10.1145/3033274.3085127. 75

[44] Goran Radanovic, Radu Jurca, and Boi Faltings, Incentives for effort in crowdsourcing using the peer truth serum, *ACM Transactions on Intelligent Systems and Technology (TIST)*, **7**(4), art. no 48, 2016. DOI: 10.1145/2856102. 76, 79

[45] Goran Radanovic and Boi Faltings, Incentive schemes for participator sensing, *Proc. of the 14th International Conference on Autonomous Agents and Multiagent Systems (AAMAS)*, 2015. 79, 80, 86

[46] Vijay Kamble, Nihar B. Shah, David Marn, Abhay Parekh, and Kannan Ramachandran, Truth serums for massively crowdsourced evaluation tasks, *arXiv:1507.07045*, 2016. 81, 82

[47] Joyce E. Berg and Thomas A. Rietz, The Iowa electronic markets: Stylized facts and open issues, *Information Markets: A New Way of Making Decisions*, pp. 142–169, 2006. 89

[48] Rafael M. Frongillo, Yiling Chen, and Ian A. Kash, Elicitation for aggregation, *arXiv:1410.0375*, 2014.

[49] Johan Ugander, Ryan Drapeau, and Carlos Guestrin, The wisdom of multiple guesses, *Proc. of the 16th ACM Conference on Economics and Computation (EC)*, pp. 643–660, 2015. DOI: 10.1145/2764468.2764529.

[50] Jacob Abernethy and Rafael M. Frongillo, A collaborative mechanism for crowdsourcing prediction problems, *NIPS*, 2011. 94

[51] Rafael M. Frongillo, Yiling Chen, and Ian A. Kash, Elicitation for aggregation, *arXiv:1410.0375*, 2015. 95

[52] Robin Hanson, Logarithmic market scoring rules for modular combinatorial information aggregation, *Journal of Prediction Markets*, 2002. 91, 92

[53] Yiling Chen and David M. Pennock, A utility framework for bounded-loss market makers, *Proc. of the 23rd Conference on Uncertainty in Artificial Intelligence (UAI)*, 2007. 91

[54] Jacob Abernethy, Sindhu Kutty, Sébastien Lahaie, and Rahul Sami, Information aggregation in exponential family markets, *Proc. of the 15th ACM Conference on Economics and Computation (EC)*, pp. 395–412, 2014. DOI: 10.1145/2600057.2602896. 92

[55] Florent Garcin and Boi Faltings, Swissnoise: Online polls with game-theoretic incentives, *Proc. of the 26th Conference on Innovative Applications of AI*, pp. 2972–2977, 2014. 51, 53, 55, 56, 93, 94

[56] Paul Resnick and Rahul Sami, The influence limiter: Provably manipulation-resistant recommender systems, *Proc. of the ACM Conference on Recommender Systems (RecSys)*, pp. 25–32, 2007. DOI: 10.1145/1297231.1297236. 98

[57] Roslan Ismail and Audun Josang, The beta reputation system, *Proc. of the 15th Bled Conference on Electronic Commerce (EC)*, 2002. 100

[58] Sonja Buchegger and Jean-Yves Le Boudec, A robust reputation system for mobile ad-hoc networks, *P2PEcon*, 2003. 100

[59] Goran Radanovic and Boi Faltings, Limiting the influence of low quality information in community sensing, *Proc. of the International Conference on Autonomous Agents and Multiagent Systems (AAMAS)*, pp. 873–881, 2016. 101, 102

[60] Nihar B. Shah and Dengyong Zhou, Double or nothing: Multiplicative incentive mechanisms for crowdsourcing, *arXiv:1408.1387*, 2015. 102

[61] Nihar B. Shah and Dengyong Zhou, No oops, you won't do it again: Mechanisms for self-correction in crowdsourcing, *Proc. of the 33rd International Conference on Machine Learning*, pp. 1–10, 2016. 102

[62] Jacob Steinhardt, Gregory Valiant, and Moses Charikar, Avoiding imposters and delinquents: Adversarial crowdsourcing and peer prediction, *arXiv:1606.05374*, 2016. 102

[63] Ofer Dekel and Ohad Shamir, Good learners for evil teachers, *Proc. of the 26th Annual International Conference on Machine Learning*, 2009. DOI: 10.1145/1553374.1553404. 105

[64] Ofer Dekel, Felix Fischer, and Ariel D. Procaccia, Incentive compatible regression learning, *Journal of Computer and System Sciences*, **76**(8), pp. 759–777, 2010. DOI: 10.1016/j.jcss.2010.03.003. 103, 104

[65] Reshef Meir, Ariel D. Procaccia, and Jeffrey S. Rosenschein, Strategyproof classification under constant hypotheses: A tale of two functions, *Proc. of the 24th AAAI Conference on Artificial Intelligence*, 2008. 103, 105

[66] Reschef Meir, Ariel D. Procaccia, and Jeffrey S. Rosenschein, Algorithms for strategyproof classification, *Artificial Intelligence*, **186**, pp. 123–156, 2008. DOI: 10.1016/j.artint.2012.03.008. 103, 105

[67] Jacob Steinhardt, Gregory Valiant, and Moses Charikar, Avoiding imposters and delinquents: Adversarial crowdsourcing and peer prediction, *NIPS*, 2016. 105

[68] Moses Charikar, Jacob Steinhardt, and Gregory Valiant, Learning from untrusted data, *STOC*, 2017. DOI: 10.1145/3055399.3055491. 105

[69] Radu Jurca and Boi Faltings, Enforcing truthful strategies in incentive compatible reputation mechanisms, *WINE: Internet and Network Economics*, pp. 268–277, 2005. DOI: 10.1007/11600930_26. 108

[70] Alice Gao, James R. Wright, and Kevin Leyton-Brown, Incentivizing evaluation via limited access to ground truth: Peer-prediction makes things worse, *arXiv:1606.07042*, 2016. 108, 110

[71] Jason Jingshi Li, Boi Faltings, Olga Saukh, David Hasenfratz, and Jan Beutel, Sensing the air we breathe—the OpenSense Zurich dataset Opensense Zurich dataset, *Proc. of the 26th AAAI Conference on Artificial Intelligence*, 2012. 111

[72] Goran Radanovic and Boi Faltings, Learning to scale payments in crowdsourcing with PropeRBoost, *Proc. of the 4th AAAI Conference on Human Computation and Crowdsourcing (HCOMP)*, 2016. 112, 113, 114

[73] Xi Alice Gao, Andrew Mao, Yiling Chen, and Ryan P. Adams, Trick or treat: Putting peer prediction to the test, *Proc. of the 15th ACM Conference on Economics and Computation (EC)*, 2014. DOI: 10.1145/2600057.2602865. 108

[74] Victor Shnayder, Rafael M. Frongillo, and David C. Parkes, Measuring performance of peer prediction mechanisms using replicator dynamics, *Proc. of the 25th International Joint Conference on Artificial Intelligence (IJCAI)*, 2016. 108

[75] Bo Waggoner, Rafael M. Frongillo, and Jacob D. Abernethy, A market framework for eliciting private data, *Advances in Neural Information Processing Systems 28 (NIPS)*, 2015. 118

[76] Arpita Ghosh, Katrina Ligett, Aaron Roth, and Grant Schoenebeck, Buying private data without verification, *Proc. of the 15th ACM Conference on Economics and computation (EC)*, pp. 931–948, 2014. DOI: 10.1145/2600057.2602902. 118

[77] Yang Liu and Yling Chen, Machine-learning aided peer prediction, *Proc. of the ACM Conference on Economics and Computation (EC)*, pp. 63–80, 2017. DOI: 10.1145/3033274.3085126. 117, 118

[78] Yang Cai, Constantinos Daskalakis, and Christos Papadimitriou, Optimum statistical estimation with strategic data sources, *JMLR: Workshop and Conference Proceedings*, **40**, pp. 1–17, 2015. 118

Authors' Biographies

BOI FALTINGS

Boi Faltings is a full professor at École Polytechnique Fédérale de Lausanne (EPFL) and has worked in AI since 1983. He is one of the pioneers on the topic of mechanisms for truthful information elicitation, with the first work dating back to 2003. He has taught AI and multi-agent systems to students at EPFL for 28 years. He is a fellow of AAAI and ECCAI and has served on program committee and editorial boards of the major conferences and journals in Artificial Intelligence.

GORAN RADANOVIC

Goran Radanovic has been a post-doctoral fellow at Harvard University since 2016. He received his Ph.D. from the Swiss Federal Institute of Technology and has worked on the topic of mechanisms for information elicitation since 2011. His work has been published mainly at AI conferences.

CPSIA information can be obtained
at www.ICGtesting.com
Printed in the USA
FSOW04n1202011117
40450FS

9 781627 057295